# Getting Started with the TI-84/83 Graphing Calculator

**Carl Swenson**

*Seattle University*

WILEY
JOHN WILEY & SONS, INC.

To order books or for customer service call 1-800-CALL-WILEY (225-5945).

ISBN 0-471-74207-4

Printed in the United States of America

10 9 8 7 6 5 4 3 2 1

Printed and bound by Hamilton Printing Company

# PREFACE

The purpose of this book is to show how to apply the features of the TI-84 and TI-83 graphing calculators to understand calculus. The book is divided into five parts, corresponding to common areas of focus in a calculus course. The chapters provide a more specific description of each calculus topic. In general, if you are looking for help on a calculus topic, then use the Table of Contents to find the topic, but if you are looking for help on a calculator command, then start by looking in the Index. Each calculus chapter is intended to be stand alone but they all require an understanding of the basics from Part I Precalculus. Part I is intended as a review; it can be skimmed by experienced users or used as a primer by new users of this calculator.

I would like to acknowledge and thank Deborah Hughes-Hallett and the Calculus Consortium for Higher Education (CCHE) for permission to use examples from their work.

## To the student

Using a graphing calculator can be both frustrating and fun. A healthy approach when you get frustrated is to step back and say, "Isn't that interesting that it doesn't work." Figuring out how things work can be fun. If you get too frustrated, then it is time to ask a friend or the instructor for help. Make sure you have a phone list of friends with the same calculator.

Part I gives you clear sets of key sequences so you become comfortable with how your calculator works. The remaining parts shift into a higher gear and only show you calculator screens as guides for the keystrokes. Your *TI Guidebook* provides a resource if you get stuck; it explains each feature briefly, usually with a key sequence example.

Remember that the *Guidebook* is like a dictionary: there is no story line or context. In this book, the features that you need for calculus are explained in the context of calculus examples. Other calculator features that are less important to calculus may not be mentioned at all. The mathematical content drives this presentation, not the calculator features.

I have included tips about such things as short-cuts, warnings, and related ideas. I hope you will find them useful.

> *Tip:* Don't use technology in place of thinking.

## To the instructor

These materials are designed to allow you to focus on the calculus, not the calculator. By having the students use a single calculator specific book, you should be able to greatly reduce the problems caused by using multiple calculator materials.

Will these materials take care of all your students? Of course not. There will still be the zealous ones who want the programs in assembly language and the anxious ones who want the buttons pressed for them. These materials are aimed at the middle, giving enough guidance so that most students are able to work through an example without assistance, but not so specific as to be considered a mindless exercise in pressing keys in the right order.

Programming is not an emphasis of this book. I have included five programs which I feel enhance the calculus learning. Find a techno-hungry student to enter them and insure that they are running properly. Then distribute them to your class using LINK.

> *Tip:*    The TI Volume Purchase Plan provides you with a classroom calculator and/or an overhead model for classroom use.

# Dedication

This book is dedicated to all golden retrievers. They know the calculus of minimizing the distance to a frisbee and maximizing the fun.

Carl Swenson
swenson@seattleu.edu

# TABLE OF CONTENTS

## PART I PRECALCULUS

# PART I  PRECALCULUS (continued)

# PART II  DIFFERENTIAL CALCULUS

# PART III  INTEGRAL CALCULUS

# PART IV  SERIES

# PART V  DIFFERENTIAL EQUATIONS

# PART V  DIFFERENTIAL EQUATIONS (continued)

**Notes:**

# CHAPTER ONE

# GETTING STARTED

Calculators have developed from the fingers, to the abacus, to the slide rule, to the scientific calculator, and now we have the graphing calculator. This chapter is a gentle introduction to using the TI-84/83 graphing calculator and shows how to make some simple numerical calculations. If you have used a graphing calculator before, you may only need to skim this chapter. The *TI–Guidebook* should also be consulted if you are having difficulty getting started. In this book the references to TI calculator keys and menu choices are written in the TI Uni font. The TI font looks like this.

## A note on different models: TI-83, TI-83 Plus, and TI-84

The oldest model, TI-83, is relatively slow and does not support the APPS feature. This feature is the ability to download, store and use special application software. An improvement, the TI-83 Plus, allows APPS, but has an inconvenient connection to a computer. TI-84 models are the fastest, have the most memory for storing APPS, and use a standard USB interface connection with computers. The TI-84 is the calculator of choice. On all three models, the keyboard and screen are similar, but not exactly the same. For example, the [2nd] key on the TI-83 is written in lower case, while on the TI-84 it is in uppercase, [2ND]. The keypads also differ in color scheme and key shape. The close similarity of keypad, screen and features makes it possible to use this book with any of the three models.

## Essential keys

### The ON key

Study the keyboard and press the ON key in the lower left-hand corner. You should see a blinking rectangular cursor. If not, then you may need to set the screen contrast. Even if the cursor is showing, it is a good idea to know how to adjust the screen contrast.

### Using the 2nd key to adjust the screen contrast

As you use the calculator, the battery wears down and it becomes necessary to adjust the screen. Also, you may need to adjust the screen contrast for different lighting environments. Press and release the 2nd key and then the up-arrow key in the upper right of the keypad. By repeating this sequence, the screen darkens. The screen can be lightened by repeating this sequence but by using the down-arrow instead of the up-arrow key. A momentary value (between 0 and 9) flashes in the upper right corner of the screen telling you the battery status

(9 is close to replacement time). If the setting is too low, the cursor does not show; if it is too high, the screen is dark as night.

If you take a break and come back later, the cursor disappears for a different reason. The calculator goes to sleep; it turns itself off after a few minutes of no activity. Just press the ON key and it wakes up at the same place it turned off: no memory loss.

> *Tip:* Sometimes "broken" calculators can be fixed by reinserting the batteries correctly.

# Arithmetic calculations on the home screen

Use technology on known results before trying complex examples. Our first use of the calculator is to do arithmetic on the home screen. Many different screens are shown later, but this *home* screen is where you do calculations. A graphing calculator has a distinct calculation advantage over a scientific calculator because it shows multiple lines and has entry recall. The numeric keys and the operation keys are used for simple arithmetic calculations.

Type some calculation for which you know the answer, say 8*9, and press the ENTER key; the result appears on the right side of the screen. You can see successive entries of three known multiplications. Had they not been seen all together, you may not have noticed an interesting pattern: the sum of the answer digits is always 9 (i.e., 7 + 2 = 9, 6 + 3 = 9, and 5 + 4 = 9). On a graphing calculator, your results flow down the screen as you work.

```
8*9
              72
7*9
              63
6*9
              54
■
```

Type: 3 ÷ 2 ENTER. Be aware that the symbol on the divide key, ÷, is different from the divide symbol (/) that appears on the screen.

Next, type: 3 – 2 ENTER. Take special care to use the subtraction key on the right side of the keypad. One of the most common errors is interchanging the use of the subtraction key – and the negation key, (−), on the bottom row.

Finally, type: 3 (−) 2 ENTER. Here the negation key creates an error. Give it a moment's thought: subtraction requires two numbers, while negation works on a single number.

```
3/2
             1.5
3-2
               1
3-2■
```

Looking carefully at the previous screen shows the two similar symbols: subtraction, which is longer and centered, and negation, which is shorter and raised. This last entry gives you a syntax error as shown here. Error screens replace the work screen and tell you briefly what is wrong. You are forced to respond with 1:Quit or 2:Goto. The Goto option is often the best choice because it will "go to" the error location and allow you to correct it.

```
ERR:SYNTAX
1■Quit
2:Goto
```

> *Tip:*    Sequences of calculator screens in an example start and stop with a heavy top and bottom border (as shown above). These borders are meant to mark the beginning and the end of a sequence; this can be especially helpful when an example extends over more than one page.

## Scientific keys

Next we test the scientific keys. These give values for many expressions used in science, such as the common log (LOG) and natural log (LN) function. The TAN, COS, and SIN keys are the standard trigonometric functions. These functions appear on the screen in lower case followed by a left parenthesis that is automatically inserted for you. First, however, we examine the taking of powers.

We know $2^3 = 8$, and to verify this on our calculator, type

2 ^ 3 ENTER

The often-used square power has the special key $x^2$. (The $x^2$ shows only $^2$ on the screen.) So pressing

5 x² ENTER

gives 25 as an answer, the same as  5^2 yields 25.

```
2^3
           8
5²
          25
5^2
          25
■
```

The effect of the $x^{-1}$ key is to take the reciprocal of a quantity, $1/x$. Thus pressing

2 x⁻¹ ENTER

gives 0.5 as an answer. To find the value of log 100, type

LOG 1 0 0 ) ENTER

Trigonometric values can also be found, as shown.

```
2⁻¹
          .5
log(100)
           2
cos(0)
           1
■
```

> *Tip*:    The right parenthesis in an expression like log(100) is not actually required, but it is a bad habit to leave it off.

## Magic tricks to change the keypad

How can we do more with the basic keys? The trick is multiple-state keys. The indicator of the keyboard state is shown by the cursor. The standard is a solid black square. The first change state key is the 2nd key in the upper left corner. After pressing 2nd once, the inside of the cursor shows an up arrow — Presto! — All the keys now have a new meaning. These meanings are indicated just above and to the left of each key, written in the color of the 2nd key. (The color is blue on a TI-84 but yellow on a TI-83.) We have already used the 2nd key to adjust the screen contrast. When the 2nd key is pressed it must be released before pressing the next key. This is not like the shift key on a keyboard, where two keys must be pressed simultaneously.

## A note about notation

Some authors use boxed text to specify calculator keys, for example, 2nd. This notation makes reading quite jarring. The TI font alone is sufficient to denote the keys, so no boxing is used in this book. When the desired keystrokes are 2nd followed by OFF, the combination will be written together as 2nd_OFF. This notation alerts you that you should look for OFF written above the key to be pressed; in this case it is above the ON.

## Practicing the 2nd key on the greatest equation ever written

Five symbols, 0, 1, e, π, and $i$, are frequently used in mathematics. Incredible as it might seem, they can be related by a single equation:

$$e^{\pi \cdot i} + 1 = 0.$$

You can practice using the 2nd key for e, π, and $i$ by entering

2nd_eˣ 2nd_π * 2nd_i ) + 1 ENTER

## Using ALPHA to store values

The other key that changes the state of keypad is the green ALPHA key; as its name indicates, it is used to enter alphabetic letters such as variables, and it has a secondary role as the Solver activate key. (This feature is discussed in Chapter 2.)

When pressed the store key, STO>, appears on the screen as: →. It is used to store a numeric value into a letter variable. If you want to repeatedly use the value of (log(100)+1)/2 in calculations, then you enter:

( LOG 100 ) + 1 ) / 2 STO> ALPHA_A ENTER

The variable A can now be used in computations, as shown.

*Tip:*   The cursor box changes if the ALPHA or 2nd keys are in effect. The ALPHA and 2nd keys work as toggles: if you press one by mistake and turn on a state, just press the key again to turn it off.

# Editing

We all make mistakes; correcting them on a graphing calculator is relatively easy. Use the arrow keys to navigate the screen and type over your errors.

## Corrections in longer expressions

The single most popular error (can errors be popular?) among new users is the failure to use parentheses when needed. In the first calculation of the previous screen, it is vital to get the parentheses in the correct places, if not the answer is different from the one shown. In general, this is a serious problem because the calculator does not stop and alert you with an error screen; instead, it gives you the correct answer to a question you are not asking. There is a prescribed order of operations on your calculator; you can look this up in your *TI–Guidebook* for details if you have questions about this order.

Suppose you want to add 2 and 6 and then divide by 8. We don't need a calculator to tell us the expression has value 1. But if you enter 2 + 6 / 8, the answer is 2.75. You can figure out that the calculator divided 6 by 8 first and then added that to 2. This was not what we wanted. We need to use parentheses to ensure that we are evaluating the correct expression. Try (2 + 6) / 8 and get an answer of 1 as expected.

```
2+6/8
                 2.75
(2+6)/8
                    1
■
```

*Tip:*    When you get an unexpected result, go back and check parentheses. Be generous; adding extra parentheses doesn't hurt.

## Correction keys: DEL, CLEAR, and 2nd_INS

While on the home screen, you can use the arrow keys to move forward and backward. (The up arrow takes you to the beginning of the line, the down arrow to the end.) Press the DEL key to delete the character in the cursor box. Use CLEAR to delete the whole entry line; if the entry line is already clear, then the whole screen clears. Thus, pressing CLEAR twice always clears the screen. If you need to insert one or more characters, you can move (using the arrow keys) to the location at which you want to insert and press the 2nd_INS key. The cursor appears as an underline when the insert mode is in effect. Pressing any arrow key turns off the insert mode.

*Tip:*    Unfortunately, there is no backspace key such as is found on computer keyboards.

# Recalling a previous entry

Often we see an error after we have pressed ENTER and left the entry line. The 2nd_ENTRY key (above ENTER) places the previous entry line on the screen so that it can be edited for the next calculation. In the last screen example, we had the expression 2+6/8 but then realized we needed to add parentheses. After seeing our mistake on the first line, we could have used 2nd_ENTRY to paste a new copy of the expression on the screen. Then, by using the navigation arrows, we could have inserted the needed parentheses. Note that we would need to use 2nd_INS since we were not writing over what was there but inserting new parentheses characters.

## Deep recall

By repeatedly pressing the 2nd_ENTRY key, many of the previous entry lines are accessible. This is called deep recall and is limited to about 128 characters. Using a previous entry overwrites the current entry; you do not go to a new line. A previous entry cannot be inserted into the command line without erasing what was already there — it is all or nothing.

## Recycling a previous answer

There is another keystroke-saving feature that is quite handy: the 2nd_ANS key. The 2nd_ANS places the variable Ans on the screen and uses the previous answer as its value when calculated.

To find the difference in area between a 10- and 12-inch pizza, we first find the 10-inch area. Next we use 2nd_ENTRY to place the 10-inch calculation back on the screen. We arrow left and change the 10 to 12.

```
π(10)^2
            314.1592654
π(1■)^2
```

Arrow back to the right, add the subtraction sign, press 2nd_ANS, and then press ENTER to find the answer.

```
π(10)^2
            314.1592654
π(12)^2-Ans
            138.2300768
■
```

If you start an expression by pressing an operation key, +, -, *, /, ^, the calculator assumes the first number in the calculation is the previous answer, so it puts Ans on the screen without your even pressing the 2nd_ANS key. Press CLEAR and then this sequence of keystrokes:

```
2                    2
Ans+3
                     5
                     8
                    11
■                   14
```

2 ENTER + 3 ENTER ENTER ENTER ENTER ENTER

*Tip:*    Work smart. Use 2nd_ENTRY and 2nd_ANS.

# Menus

We have already seen one example of a menu when we committed the error of using the negation sign in place of the subtraction symbol. There were only two choices, but the principle is the same in other cases. Menus are lists of choices that are far too numerous to be available on the keypad.

Press CLEAR twice to start with a fresh screen. Press MATH and you see a menu screen that takes the place of the home screen. Notice that the options are numbered and you can choose by number or by arrowing to a choice and pressing ENTER. All the selections may mean little to you at this point at this point.

```
MATH NUM CPX PRB
1▮►Frac
2:►Dec
3:³
4:³√(
5:ˣ√
6:fMin(
7↓fMax(
```

The top row shows four menus: MATH NUM CPX PRB. Press the right-arrow key to get the NUM menu. The abs( is our choice, so we press ENTER (or 1).

```
MATH NUM CPX PRB
1▮abs(
2:round(
3:iPart(
4:fPart(
5:int(
6:min(
7↓max(
```

Now the home screen reappears and abs( has been pasted onto the home screen. If the home screen is not cleared, the pasting takes place at the location of the cursor, even in the middle of any expression.

```
abs(█
```

In case you hadn't guessed, abs() is the absolute value function, which makes numbers positive. Here we see it work. Note that we used the negation sign, not the subtraction key. (Using the subtraction key causes an error.)

```
abs(-6)
              6
█
```

*Tip:*    Most accomplished users use the number choice as the fastest method of choice in menus. However, if the choice is the first item, then ENTER is best.

*Tip:*    A menu may not show all its items. An arrow in place of the colon indicates that unseen items are available in the arrow's direction.

*Tip:*    Menus are wrap-around so you can press the up arrow once to get to the last item in a list. A choice of menus on the top line wraps from left or right by using the left or right arrows.

# A CATALOG of items

An alphabetical list of feature is available. The 2nd_CATALOG key shows an alphabetic list of items. The list is long, but the black boxed A, in the upper right corner is a reminder that you are in ALPHA mode and you can conveniently jump closer to the desired item by pressing the key of items first letter.

```
CATALOG        A
▶abs(
 and
 angle(
 ANOVA(
 Ans
 Archive
 Asm(
```

# Changing the format: MODE

You can control the format of the output for your numerical calculations For example, if you are doing a business application, you might want money answers to come out rounded to two decimal places for the dollar-and-cents format. The MODE screens shown are from the TI-84. The TI-83 screen is the same except for the SET CLOCK feature on the bottom line. This time and date feature has been added to the TI-84.

The MODE key allows you to check and change formats. The default settings are all on the left of the screen so a quick glance tells you if any settings have been changed. Numeric answers are usually shown in FLOAT mode, meaning they show up to eight decimal places if needed.

```
NORMAL  SCI  ENG
FLOAT  0123456789
RADIAN  DEGREE
FUNC  PAR  POL  SEQ
CONNECTED  DOT
SEQUENTIAL  SIMUL
REAL  a+bi  re^θi
FULL  HORIZ  G-T
            04/10/11 18:37
```

To change the setting to show dollar-and-cents, use the down arrow to reach the FLOAT line and then use the right arrow to move across to the desired setting. You must now press ENTER to make the change. Press 2nd_QUIT to return to your home screen.

```
NORMAL  SCI  ENG
FLOAT  01■3456789
RADIAN  DEGREE
FUNC  PAR  POL  SEQ
CONNECTED  DOT
SEQUENTIAL  SIMUL
REAL  a+bi  re^θi
FULL  HORIZ  G-T
            04/10/11 18:38
```

We calculate the total cost of a $2.10 Cafe Latte in Seattle where the sales tax is 8.6%. Since we have changed the format to two decimals, our answer is rounded to 2.28 Rounding the result in a display format does not change the stored accuracy. To see this, multiply the answer by 100 and you see that the fuller decimal accuracy is preserved.

```
2.10*1.086
                 2.28
Ans*100
               228.06
■
```

For calculus, the RADIAN angle setting is used. This is the default on the MODE menu. For situations where angles in degrees are specified, use the 2nd_ANGLE menu to paste the degree symbol into the calculation. This is preferable to changing the MODE setting to DEGREE.

```
ANGLE
1:°
2:'
3:r
4:▶DMS
5:R▶Pr(
6:R▶Pθ(
7↓P▶Rx(
```

Trigonometric calculations are normally in radian measure. Use the 2nd_ANGLE menu to specify degree or radian measure if needed. Check the entry of angles using a sine function.

```
sin(30)
          -.9880316241
sin(30°)
                    .5
sin(30r)
          -.9880316241
■
```

The MODE setting choices are described in the *TI–Guidebook*. We mention other settings as we need them, but unless noted otherwise, all our examples assume that the default settings are in effect.

*Tip:*   If your output values are in an unexpected or undesirable format, check the MODE settings. If you are having trouble changing the MODE settings, you may have forgotten to press ENTER before 2nd_QUIT.

# CHAPTER TWO

# DEFINING FUNCTIONS

The definition of functions and the use of functional notation are vital to success in calculus. In the next three chapters, we use the TI calculator's Graph/Table keys to define and evaluate functions, to make tables of values, and to graph functions. In short, we will see how to view functions analytically, numerically and graphically. In this chapter, we focus on the first key, Y=, of the top row of Graph/Table keys.

## Formula vs. function notation

Function notation is used in calculus, whereas formulas are used in algebra. So what is the difference and how are they related? They both express a relationship between variables. Let's take the famous formula for the area of a circle, $A = \pi r^2$. In precalculus you learned to write this in functional notation $f(r) = \pi r^2$. The functional notation tells you <u>explicitly</u> what variable is the independent variable.

You can define up to ten functions in the function editor; the editing screen appears when you press the Y= key. The ten available functions are labeled $Y_1$, $Y_2$, ... ,$Y_9$, $Y_0$. You use the up and down arrow keys to scroll and see them all. To define a function, it is easiest to think of it in formula form. Choose one of the ten Y to be the dependent variable and make X the independent variable. For example, in our circle formula we can define $Y_1=\pi X^2$, where $r$, the independent variable, is replaced by X. See Figure 2.1 for this and other examples.

The independent variable has a special key, $X,T,\theta,n$, which writes the variable X without using the ALPHA key. (The other symbols $T,\theta,n$ on this key are active in different modes that are set in the MODE menu, but these do not concern us at this point.)

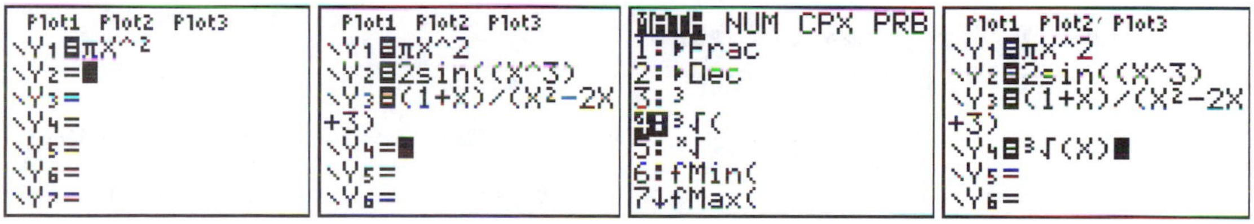

*Figure 2.1 Defining functions from the keypad and pasting from the MATH menu.*

### Pasting from a menu

There are functions, such as the cube root, that can be pasted into a function definition from a menu, such as the MATH menu. Press MATH, use the down-arrow to $4:\sqrt[3]{\ }($, and press ENTER to paste the notation in place at $Y_4$. See Figure 2.1.

*Tip:*    The independent variable must always be called X.

## Pasting from the CATALOG

Remembering where special symbols and functions are located within various menus can be tedious. The most convenient way to paste a non-keypad expression into a function definition is by using 2nd_CATALOG. In the catalog screen you can move quickly to the function you want by pressing the letter key that starts its name. For example, if you want to use the hyperbolic tangent (tanh), then press 2nd_CATALOG and you will see the list beginning with abs(. Now press T (no need to press ALPHA because the upper right indicator shows it is in Alpha mode). Finally use the down arrow to highlight the entry tanh( and then press ENTER.

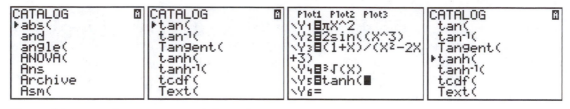

Figure 2.2  Using the CATALOG to define a function.

## Cleaning up and getting out: QUIT

Use the arrow keys to navigate on the function editor screen. Use DEL and INS to edit. Pressing CLEAR to delete any definition. Use 2nd_QUIT to return to the home screen.

> **Tip:**  Be careful about using the CLEAR key. It instantly deletes the entire entry and there is no recovery other than re-entering the formula.

# Evaluating a function at a point

A benefit of functional notation, *f(x)*, is that *f(3)* is conveniently understood to be the output value of the function when 3 is input. Since $Y_i$ stands for the function $Y_i(X)$, we evaluate it at points where it is defined, for example, $Y_1(3)$. To find the area of a circle with radius 10 cm, we will use the area function that was defined as our first function. (Enter it again if you have deleted it.)

### Getting Y from the VARS menu

To find the value of $Y_1(10)$ from the home screen, press

VARS    Arrow-Right(selects Y-VARS)    ENTER(selects 1:Function…)    1(selects Y1)

(ENTER would also select $Y_1$, but you would need to type a number or use the arrow keys to select a different function), then back on the home screen add: ( 1 Ø ) ENTER

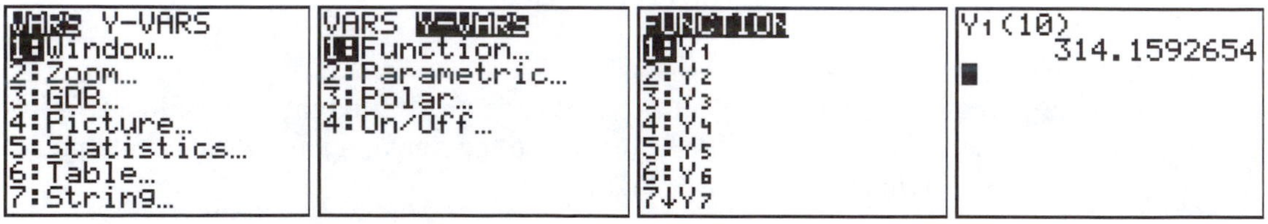

Figure 2.3  Evaluating $Y_1$ using the VARS,  Y-VARS,  1:Function… sequence.

> **Tip:** You must use VARS to access $Y_1$ — you cannot type Y 1 (this would be Y∗1).

> **Tip:** The key sequence  VARS , Y-VARS , 1:Function… will be used repeatedly.

# New functions from old

In the next two examples, we create new functions from previously defined ones.

### Composite functions

Suppose an oil spill expands in a perfect circle and that the radius increases as a linear function of time. We create a new composite function that expresses the area in terms of time. Let $f(r) = \pi r^2$ and $g(t) = 1+t$, where $t$ is in hours. Our new function is $h(t) = f(g(t)) = \pi(1+t)^2$. In Figure 2.4 we create this composite function and find the area of the oil spill after 2 hours. Notice that $Y_3(2)$ gives the same answer as $Y_1(Y_2(2))$. Even in composite functions all the independent variables must be entered as X.

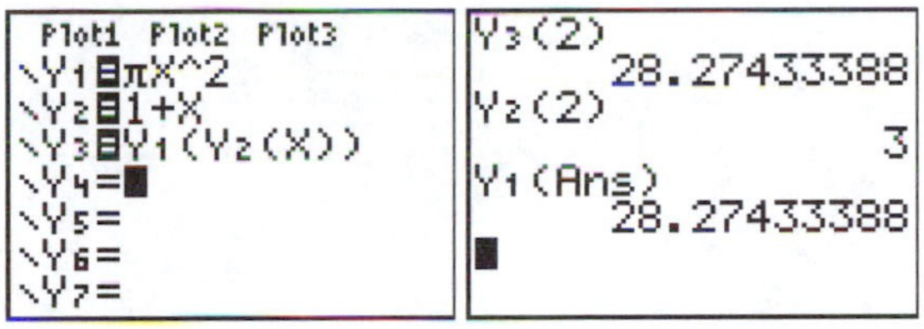

```
Plot1 Plot2 Plot3      Y3(2)
\Y1◼πX^2                       28.27433388
\Y2◼1+X               Y2(2)
\Y3◼Y1(Y2(X))                            3
\Y4=◼                 Y1(Ans)
\Y5=                          28.27433388
\Y6=                  ◼
\Y7=
```

*Figure 2.4  Creating a composite function for area in terms of time.*

### A Malthusian example

In 1798, Thomas Malthus proposed that population growth was exponential and that food supply would grow at a linear rate. We model a food supply (per millions of people) as $Y_1 = 5 + .2X$; this means that there is food for five million people in the base year and that each year the supply increases to provide for an additional 200,000 people. (As you enter these functions, the previous ones are erased.) For the population (in millions), set $Y_2 = 2(1.03)^X$; this means that the population starts at two million and increases annually by three percent. Let $Y_3 = Y_1 - Y_2$ and $Y_4 = Y_1 / Y_2$. These two new functions are a measure of excess food and a measure of food per capita, respectively. In Figure 2.5 these two measures are evaluated at 50 and 100 years from the base year. There is a shortage in the 100th year ($Y_3$ is negative and $Y_4$ is less than 1). To find the first year of shortage we need a graph or table. This is done in the next chapters.

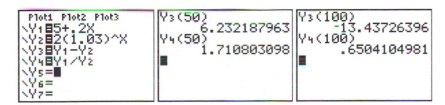

*Figure 2.5  Food excess and food per capita in 50 and 100 years.*

# Defining families of functions by using lists (Optional)

Let's get tricky and see how to define a family of linear functions in just one single function definition. Suppose a city has three taxi companies. Red charges $1.00 to get in one of its taxis and $0.40 for each eighth of a mile traveled. Green charges $2.00 to get in and $0.30 for each eighth of a mile traveled. Blue charges $3.00 to get in and $0.20 for each eighth of a mile traveled. The cost in terms of miles traveled is linear; for example, Red has the cost function $C = 1+3.2x$ (for $x$ in whole miles). Since all three are linear functions of the same form, ($C = b+ax$), we use just one function definition whose parameters, $a$ and $b$, are lists. To enter this example use the list (or set) symbols: { and }; they are above the parenthesis keys in the 2nd mode.

Figure 2.6 shows the charges for one and two mile trips, the outputs are lists. We see that Red is cheapest for a 1 mile trip but Blue is cheapest for a 2 mile trip. It should be clear from the rate structure that for longer trips, Blue is cheapest.

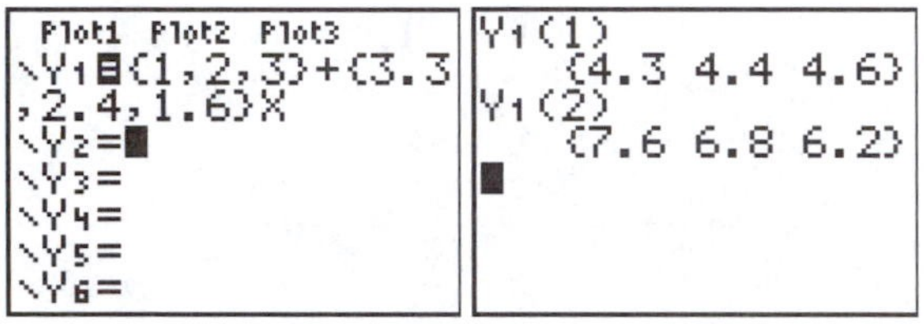

*Figure 2.6  A family of functions created using a list.*

# A short summary of menu use

A menu fills the entire screen. Menu names appear on the top row of the screen in all caps. Menu selections are numbered and usually include some lower case letters. You use the up/down arrow keys to navigate to a selection and press ENTER to make the selection. If there are more than seven selections in a menu, then an arrow is shown on the bottom line to indicate that more selections can be seen by scrolling. As you become more familiar with the choices of numbered menus, you can streamline your election process by just pressing the desired selection number.

When the top line of a menu has more than one entry, like VARS and Y-VARS in Figure 2.3, there are multiple menus. Use the left and right arrows to switch among multiple menus.

There are two kinds of menu items, those with an ellipsis (…) and those without. Items without ellipsis paste commands at the cursor position or start a process. Those items with an ellipsis show a dialog screen that often is another menu or sometimes a screen with a group of settings to be edited.

# CHAPTER THREE

# MAKING TABLES OF FUNCTION VALUES

This chapter's focus is making tables of function values. These values often reveal the nature of the functional relationship: are the values increasing?, decreasing?, periodic?

## Lists of function values

To see a set of values for a function, evaluate the function with a list as the input variable and then see the list of corresponding output values. We used a set in the definition of a function in the last chapter for the taxi company example. Here we use a set as the input variable. For this example, suppose you want to find the area of circles with radii 10 cm, 50 cm, and 100 cm.

- Enter the area function in Y₁ and type 2nd_QUIT to return to the home screen.
- Evaluate the function with a list as the input: Y₁({10,50,100}) (recall that the Y₁ is pasted from the VARS menu and that the list symbols, { }, are needed to create a list).
- When data is too long to fit on the screen, use the left/right arrow keys to scroll horizontally. Look for "..." at the right or left of the screen to indicate that there are more values off the screen in that direction.

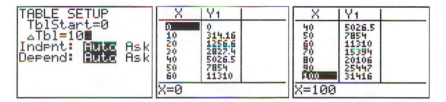

*Figure 3.1 Values of the area function at 10, 50, and 100. Scrolling to see whole list.*

## A table of values for a function

There is a much more convenient way to see a list of value for a function. We simply enter the function like before, set the beginning and increment for a table, and display it. In Figure 3.2, let's set up a table using 2nd_TBLSET (on the Graph/ Table row). We redo the above

*Figure 3.2 Making a table for the area function then using the down arrow to scroll to a value for x=100.*

problem by using a table beginning at zero (TblStart=0) and having the X values go up by ten (ΔTbl=10). Now press 2nd_TABLE (also on the top row) to display the table. In the first table of Figure 3.2 the value for 100 is not shown, but we can use the down arrow to scroll and see its value in the table. New table values are calculated as you scroll.

### Selecting special values for a table

Suppose you want the function evaluated at the three specific values 10, 50 and 100. In the TBLSET screen, we can arrow down to the option Indpnt: Auto and use the right arrow to change from Auto to Ask. Now press ENTER to set the change before you press 2nd_QUIT. The calculator ignores the TblStart and ΔTbl values and allows you to manually enter the specific values you want to evaluate. See Figure 3.3.

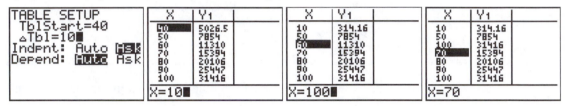

*Figure 3.3 How to enter selected values rather than an incremented list.*

# A table for multiple functions: table scrolling

Recall that in the Malthus models from the previous chapter we had a food supply function and a population function. If we want to enter these function definitions again but also want to keep our area function as $Y_1$, then we use $Y_2$ and $Y_3$; see Figure 3.4. If you reset the TBLSET screen to TblStart=0 and ΔTbl=10 and then press TABLE, then you see the $Y_1$ values, which we don't care about right now. Press the right arrow and scroll over to see the $Y_2$ and $Y_3$ values. Notice that the column for the X values remains on the screen.

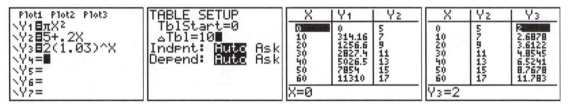

*Figure 3.4 Using right/left arrow keys to scroll and see selected columns of values. The X column remains stationary.*

### Selecting and Deselecting a function

When you want to see values for only certain functions, you can deselect the ones you do not want and select the ones you do want. On the Y= screen, move the cursor to the equal sign of the function definition and press ENTER. This is a toggle: if it was on (denoted ▊), it turns off

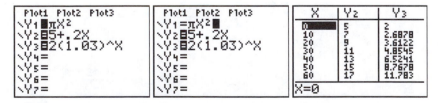

*Figure 3.5 Selecting and deselecting a function by pressing ENTER while the cursor is on its = sign.*

(denoted =); if it was off, it turns on. When you make a function definition, it is automatically turned on. With Y₁ deselected, as shown in Figure 3.5, only Y₂ and Y₃ show in the table display.

## Find the zero of a function from a table

A natural question arises from the Malthus model: When will the food supply no longer be sufficient for the population? In the previous chapter, we had a function that measured the excess food supply, so we want to find the years for which the excess food function gives negative values. However, it is easier to start by looking for when the excess food function is zero. Enter the relevant functions again and turn off any functions that we don't want to see. Since we know from before that the excess is positive at year 50 and is negative by year 100, we set TblStart at 50 and ΔTbl at 10 (see Figure 3.6). We see that the zero is between year 70 and 80. Now use 2nd_TBLSET again with TblStart at 70 and ΔTbl at 1. You need to arrow down to see that the zero is between 79 and 80. We could continue searching for a more precise value by setting TblStart=79 and ΔTbl=0.1.

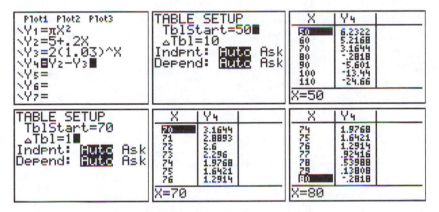

Figure 3.6 Searching for a zero of the food excess function.

*Tip:*    Deselected functions are still active for calculations when used in other function definitions.

## Editing a function formula from inside a table

You can redefine a function from the table itself. Suppose you want to look at the food ratio function in the Malthus model. Arrow up to the Y₄ cell and its equation is shown on the bottom line of the screen. Press ENTER to begin editing, arrow right and replace - with ÷, then press ENTER to finish editing. The cursor moves down from Y₄ to the new values.

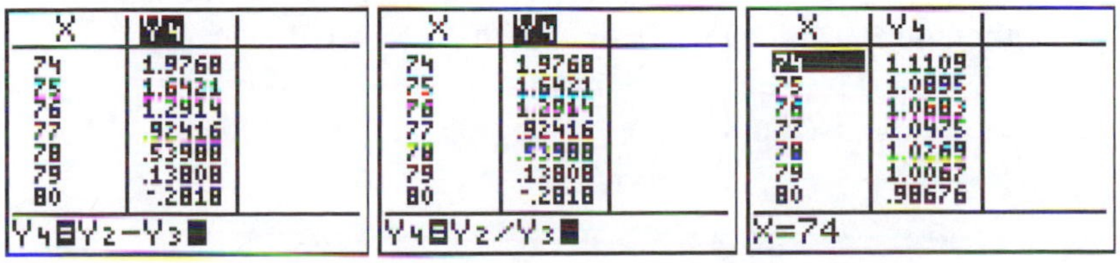

Figure 3.7 Changing a function definition from inside a table.

# CHAPTER FOUR

# GRAPHING FUNCTIONS

This chapter completes our investigation of the Graph/Table keys and shows how to graph the functions that we have defined.

## Basic graphing: WINDOW and GRAPH

Graphing is like the 1-2-3 of taking a picture with a camera.
1.  *Select your subject(s)*. To select a function, recall that you highlight its equal sign on the Y= screen (so it appears as ▊). (Turn off or clear functions that you do not want to graph.)
2.  *Frame them properly*. Press WINDOW and set the *x* and *y* window boundaries.
3.  *Click to take the picture*. Press GRAPH.

The hard part of photography is getting the subject both in the picture and looking good. On the calculator we control the picture by using the WINDOW menu. In Figure 4.1 the first row shows the settings and graph for a picture of the function $Y_1 = \pi X^2$. The function's graph uses too little of the screen. The last frame in Figure 4.1 shows an improvement made by setting Ymax=300 and Yscl=100 in the WINDOW screen.

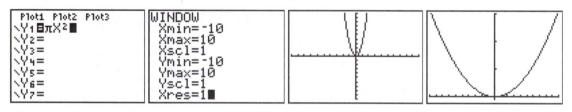

*Figure 4.1 The basic graph sequence: Y=, WINDOW, GRAPH and then an improvement by changing the WINDOW settings.*

### Window settings: Xmin, Xmax, Xscl, Ymin, Ymax, Yscl, Xres

The window setting variables are as follows:

Xmin  sets the left edge of the window as measured on the horizontal axis,

Xmax  sets the right edge of the window as measured on the horizontal axis,

Xscl  (X scale) sets the width between tick marks on the horizontal axis,

Ymin  sets the bottom edge of the window as measured on the vertical axis,

Ymax  sets the top edge of the window as measured on the vertical axis,

Yscl  sets the width between tick marks on the vertical axis, and

Xres  (X resolution) sets the selection density of values to plot (1 is highest setting and should be used unless the graphing is very slow).

# Using a preset window: ZOOM

Like photography, the setting of the window is an art. It is rare for us to know the ideal window before we graph; trial-and-error experimentation is usually required. To help with this process, use the ZOOM menu to get started. Press ZOOM, the middle key of the Graph/Table row. Use the down arrow key to see all the choices. The item MEMORY on the top line indicates a second menu. This menu (third frame of Figure 4.2) is seen by using the right arrow key. The 1:ZPrevious option returns you to the previous settings which is handy if you change the window settings and the graph is worse. The second and third items in this menu allow you to keep a window setting in memory. The fourth setting allows you to change the size of the multiplier/divider for zooming. The default setting is 4, which is good for our use.

*Figure 4.2 The complete ZOOM screen. To see it all requires scrolling.*

### Zoom special settings: ZStandard, ZDecimal, ZTrig

There are three ZOOM menu options that automatically – all in one keystroke – set the window to special settings and graph selected functions. These special settings are shown in Figure 4.3. They are especially helpful when you are graphing common functions with a graph close to the origin. The 6:ZStandard often works well as a good first view. The 4:ZDecimal is called a *nice* window because the $x$-values used to graph progress from -4.7 to 4.7 by tenths. The nicety of this is explained with the TRACE feature just ahead. For trigonometric functions, the obvious first choice for graphing is 7:ZTrig; it uses $x$ values from -2π to 2π (in decimal form).

*Figure 4.3 ZStandard, ZDecimal, and ZTrig window settings.*

### Window adjustment for Malthus: ZFit

When graphing functions that model a situation, you almost always know the domain of the function but probably not the range. If you enter the domain, then use Ø:ZFit, often sets good $y$-values. Let's try this out on the Malthus model of the previous chapters.

Malthus never published a graph; he used only numerical and analytical expositions. Some of his readers didn't see his concern. A graph appeals to a broader audience. We use Ø:ZFit find a good window for the Malthus graphs. Enter or turn on the two Malthus equations. Use WINDOW to set Xmin=Ø and Xmax=1ØØ, as we look over a century of time. The range of the functions is difficult to guess, so ignore the Y settings and use ZOOM Ø:ZFit to set the $y$-axis window settings and graph the function as shown in Figure 4.4. (The $x$-axis window settings are unchanged by Ø:ZFit.) To see the values that have been set, press WINDOW. From this point, you may want to make minor adjustments, such as setting the Xscl and Yscl, to improve the graph.

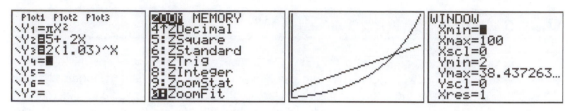

*Figure 4.4 Set Xmin=0 and Xmax=100 before using ZFit to find the range for the Malthus functions.*

> ***Tip:***   Set Xscl and Yscl to zero as you adjust a window, then choose helpful values for them when the final window size is found.

> ***Tip:***   The menus are wrap-around, so that you can get to 0:ZFit by pressing the up arrow just once.

Since there are no numeric labels on the graph, it is often necessary to alternately use the GRAPH and WINDOW keys to check on the window dimensions. This problem of knowing where you are on an unmarked graph is solved by the TRACE feature.

### Other ZOOM options: ZBox, ZSquare, ZoomStat

For the sake of completeness, we mention other choices on the ZOOM menu.

The 1:ZBox option is similar to 2:Zoom In and 3:Zoom Out in that it displays the small cross-hair cursor on the graph. Move it to the screen location where you want a new window to have one corner, press ENTER, then move the cursor to the diagonal corner desired (you see the rectangle on the screen as you move the cursor), and press ENTER again. The new window is the rectangle you defined.

The 5:ZSquare selection is helpful for graphing circles. It is similar to 6:ZStandard in that it is a one-keystroke grapher, but the window settings are changed depending upon the current settings: it adjusts the $y$ settings so that the $y$-axis has the same scale as the $x$-axis.

The 9:ZoomStat entry on the ZOOM menu graphs a good window for statistical data. We use this only once in a later chapter.

## Identifying points on the screen

By pressing any arrow key with a graph screen showing, a cross-hair cursor appears in the center of the screen and the $x$- and $y$-values of the cursor point are displayed. This cursor is called the free-moving graph cursor. (Since we are discussing graphs, we just call it the free-moving cursor.) You use the arrow keys to navigate this cursor to any point on the screen. By placing the free-moving cursor on the graph of a function, you display the approximate coordinate values of the point $(x, f(x))$. If you press CLEAR, the free-moving cursor mode is canceled.

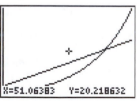

*Figure 4.5 The free-moving cursor, activated by an arrow key and deactivated by CLEAR.*

### Reading function values from the graph: TRACE

You are usually more interested in the values of a function than general points on the screen. We improve on the free-moving cursor approach by using the TRACE mode. The TRACE key is fourth in the Graph/Table row and it produces the trace cursor, a box with a blinking ×. The trace mode displays the *x*- and *y*-values of the current trace cursor location on the bottom row and the function formula on the top of the screen. Use the left and right arrows to move to other points on the graph as shown in the second panel of Figure 4.6. The third panel of Figure 4.6 shows the trace cursor being switched to the other graph: the up and down arrows select the function to be traced. If you press CLEAR, the TRACE mode is canceled.

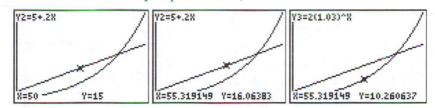

Figure 4.6 Using arrow keys to move the *TRACE* cursor on a graph.

> **Tip:** Even if you are not displaying the graph screen, the TRACE key graphs and displays the trace cursor. So a natural key sequence for graphing is Y=, WINDOW, and TRACE.

### How to make nice windows (an optional adventure)

The screen is 95 pixels wide and this is why ZDecimal is so nice: with Xmin=-4.7 and Xmax=4.7, the resulting *x*-values are 47 negative tenths, zero, and 47 positive tenths, a total of 95 pixels. For the Malthus graph, a nice window would be Xmin=0 and Xmax=94. In general, for a graph starting at Xmin=0, set Xmax=9.4*n, where *n* is large enough to let 9.4*n span the *x*-values you want to include. This produces a nice window!

In Figure 4.6, while tracing, you saw the *x*-value 55.319149. If we just wanted a rough idea about values, we could ignore the extra decimal digits. A better option is to use ZOOM 8:ZInteger which resets the window so that the trace values are integers. The 8:ZInteger selection displays a small cross-hair cursor for you to place as the center of the graph. In this case, it is centered already, so we just press ENTER. Use TRACE to see that the *x*-values are now integers.

# Panning a window

There is an old story about blindfolded people describing an elephant from different perspectives; their guesses included wall, tree, and snake. Sometimes functions are like elephants: you need to take the blindfold off to see the whole picture. In case the subject is too big to fit in one window, we move the window frame to see what is to the left or the right or above or below the current view. This is called panning. Let's take the example of the graph of a logistic equation and start as if we knew nothing about it.

$$f(x) = \frac{1000}{1 + 9e^{-0.05x}}.$$

Enter the function and get a first look by using ZOOM 6:ZStandard. We see no graph. This is the $-10 \leq x \leq 10$ and $-10 \leq y \leq 10$ window.

```
Plot1 Plot2 Plot3
\Y1◼1000/(1+9e^(
-.05X))
\Y2=
\Y3=◼
\Y4=
\Y5=
\Y6=
```

Press TRACE to find <u>some</u> x, y value. Then press ENTER to 'Quick Zoom'. This pans the window and places the trace cursor at the center. Quick Zoom always brings some part of the graph into the window.

```
Y1=1000/(1+9e^(-.05X))
X=0       Y=100
```
```
Y1=1000/(1+9e^(-.05X))
X=0       Y=100
```

The Quick Zoom gave too close a view. Instead, let's try setting the x-values over a broader interval, say $0 \leq x \leq 100$, and then using ZFit to see a more complete graph.

```
WINDOW
 Xmin=0
 Xmax=100◼
 Xscl=1
 Ymin=90
 Ymax=110
 Yscl=1
 Xres=1
```

To explore at a specific point, like $x = 100$, we press TRACE 1 Ø Ø which evaluates the function and displays the trace cursor there.

```
Y1=1000/(1+9e^(-.05X))
X=100◼
```
```
Y1=1000/(1+9e^(-.05X))
X=100       Y=942.82562
```

As before, pressing ENTER, while in the TRACE mode, pans to a graph centered at $x = 100$. To investigate further to the right, hold down the right arrow key: when the trace cursor reaches the right side of the screen, the screen pans to the right.

```
Y1=1000/(1+9e^(-.05X))
X=100       Y=942.82562
```
```
Y1=1000/(1+9e^(-.05X))
X=153.19149 Y=995.77436
```

After seeing various aspects of the function, press WINDOW, enter a good window, and press TRACE.

```
WINDOW
 Xmin=0
 Xmax=200
 Xscl=50
 Ymin=0
 Ymax=1200
 Yscl=100◼
 Xres=1
```
```
Y1=1000/(1+9e^(-.05X))
X=100       Y=942.82562
```

In summary, we used two techniques to move a window without giving specific numbers. The Quick Zoom command is TRACE ENTER, which pans and centers the window on the trace cursor. Secondly, the trace cursor, when moved to the left of the Xmin value or to the right of the Xmax value, pans the window horizontally.

*Tip:*    Pressing 2nd_ right or left arrow puts the trace cursor in *turbo* mode — it moves faster.

*Tip:*    The free-moving cursor does not pan the screen.

# Finding a good window

You saw the trial-and-error approach to finding a good window, but this begs the question: what is a good window? For a function serving as a model of a physical situation, a good window shows the function's graph for the relevant domain. For example, there is no interest in negative values of the area function $A = f(x) = \pi x^2$. But considering the same function as a purely algebraic quadratic function, we would choose a window that includes negative $x$-values to see more of its general behavior. Whenever possible, we want to show asymptotic behavior. For example, we needed to see past $x = 100$ in the graph of the logistic function because that function is asymptotic to the line $y = 1000$. We call this the end-behavior. However, if we concentrate solely on the end behavior, we might blur some local behavior. In the logistic example, an important local behavior is when the graph changes from being concave up to being concave down.

### Can you always find a good window?

No. There are pathological functions that we cannot graph and others that require more than one view to see all the important behavior. Sometimes the end-behavior view makes it impossible to see the local behavior and vice versa.

### Asymptotic dangers: Beware of graphs with vertical lines

Sometimes the graphing calculator leads you astray. The most common case is rational functions. Let's take the blind graph approach to

$$p(x) = \frac{x^2 + 2x + 30}{x - 4}$$

Don't forget to use parenthesis for the numerator and denominator as you enter the function, Y₁=(X²+2X+3Ø)/(X-4).

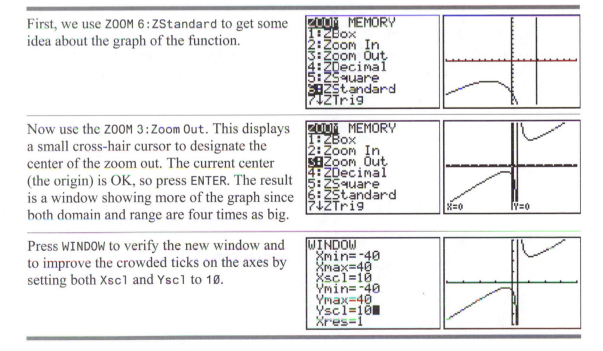

First, we use ZOOM 6:ZStandard to get some idea about the graph of the function.

Now use the ZOOM 3:Zoom Out. This displays a small cross-hair cursor to designate the center of the zoom out. The current center (the origin) is OK, so press ENTER. The result is a window showing more of the graph since both domain and range are four times as big.

Press WINDOW to verify the new window and to improve the crowded ticks on the axes by setting both Xscl and Yscl to 1Ø.

> *Tip:*    When at the origin, the small cross-hair cursor is hard to see since it appears as a single blinking pixel.

In the last sequence we are left with an unexplained vertical line to the right of the origin. Could this be part of the graph? A careful look at the function shows that the denominator is undefined at $x = 4$. Recall from precalculus that such a line is called an asymptote. But the calculator did not draw this as an asymptote. The calculator draws graphs by connecting special $x$-values that are found by starting at `Xmin` and adding increments of (`Xmax` - `Xmin`)/95. In this case, to the left of $x = 4$, $f(3.4042553) = -812386$, and to the right of $x = 4$, $f(4.2553192) = 221.75532$. Connecting these values gave us the vertical line.

### Changing plot style in `MODE`: `CONNECTED` to `DOT`

The graph connecting problem can be remedied by changing the plot style from `Connected` to `Dot` on the fifth line of the `MODE` menu as shown in Figure 4.7. Later we see how to change individual plot styles when needed.

A second approach to stop the connection across an undefined point is to create window settings with the undefined value of $x$ exactly in the middle of `Xmin` and `Xmax`. See Figure 4.8. The trace cursor shows that the value is undefined at $x = 4$ by not showing a $y$-value.

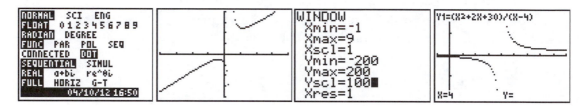

*Figure 4.7 Changing a function format to dot display.*

*Figure 4.8 Window setting so that the trace cursor shows the undefined point at x = 4 and no y value.*

### An inaccurate graph

A rational function graphed on a calculator may connect dots when it should not. A similar distortion occurs when the resolution is insufficient to display a function. Consider the function

$$f(x) = \sin\left(\frac{1}{x}\right)$$

There is no way this function can be accurately graphed if the origin is shown.

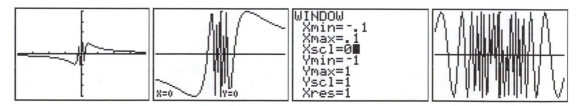

*Figure 4.9 Attempts to graph y = sin(1/x) using* `ZTrig`, `Zoom In`, *and the window -0.1 ≤ x ≤ 0.1.*

# Graphing inverse functions

The graph of $f^{-1}$ is the reflection of the graph of $f$ about the line $y = x$. This is seen in Figure 4.10 for the graph of the function $f(x) = x^3$.

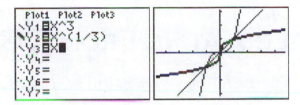

*Figure 4.10 The graph of f and $f^{-1}$ as a reflection about the line y=x. (ZDecimal window)*

In the previous example, the inverse function graph is shown in bold style. The slash sign that you have seen to the left of the Y= is the icon for a regular line. Select a graph style in the Y= edit menu, by arrowing left to this icon, you can change it to a bold line by pressing ENTER. Unfortunately, to change back to the regular style, you must press ENTER six times to cycle through the entire set of styles.

We now graph $f(x) = \sin(x)$ and its inverse, $\sin^{-1}(x)$, in the same window. We see in Figure 4.11 that some of the graph's reflection across the $y = x$ line is missing. The difference from the previous example with the cubic function is that the reflection of the sine is not a function. The $\sin^{-1}(x)$ function has to be restricted to make it a function.

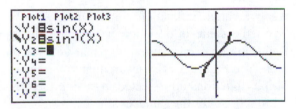

*Figure 4.11 A function and its inverse where the inverse is not a complete reflection across the line y = x. (ZDecimal window)*

To see the full reflection select only Y₂, and then on the home screen use 2nd_DRAW 8:DrawInv and paste Y₁, as shown in Figure 4.12. The Y₂ graph can be traced but not the lighter drawn inverse.

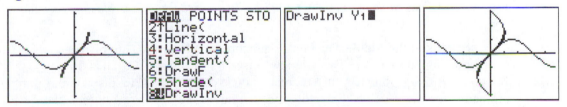

*Figure 4.12 The DrawInv command draws a complete reflection even thought the graph is not a function.*

# CHAPTER FIVE

# CALCULATING FROM A GRAPH

We have used the five basic Graph/Table keys on the top row of the calculator:

Y=    WINDOW    ZOOM    TRACE    GRAPH

and two of 2nd keys for tables:

2nd_TBLSET    2nd_TABLE

This chapter discusses the remaining three 2nd Graph/Table keys. The main focus is the 2nd_CALC menu which helps locate special points and values on a graph. The remaining two menu keys: 2nd_FORMAT and 2nd_STAT PLOT are mentioned in closing to complete our tour of the Graph/Table keys.

## Finding special values on a graph: 2nd_CALC

### Evaluating a function

In Chapter 2 we evaluated a function for a given value of $x$. The 2nd_CALC command 1:value offers you the same information by an alternate method. However, it includes a visual bonus of marking an × on the graph so that you see the value in relationship to other points on the graph. See Figure 5.1. Be aware that you must request values between Xmin and Xmax. When multiple functions are selected, by using the down arrow, you see the values of the other selected functions for the same X value.

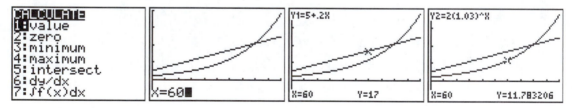

Figure 5.1 Malthus equations in $0 \leq x \leq 100$ and $-10 \leq y \leq 40$ window. Evaluating each function at x=60.

Let's look at the Malthus example one more time. The food supply function is $Y_1$ and the population function is $Y_2$. The window is set for $0 \leq x \leq 100$ and $-10 \leq y \leq 40$. Pressing 2nd_CALC displays the CALCULATE menu. Press ENTER or 1 to display the graph and be prompted for the $x$-value. Enter 60 and to find Y1(60) = 17. You can then evaluate the second function, $Y_2$, by pressing the down arrow.

***Tip:*** The special points of the CALC menu can only be found between the current Xmin and Xmax.

### Finding zeros

The second CALC tool, 2:zero, helps us find the zero (or root) of a function.

We continue with the Malthus model and consider the food excess function ($Y_3 = Y_1 - Y_2$; remember that you must use VARS to enter this). The zero of this function has an important meaning; it is the point where we start having a food shortage.

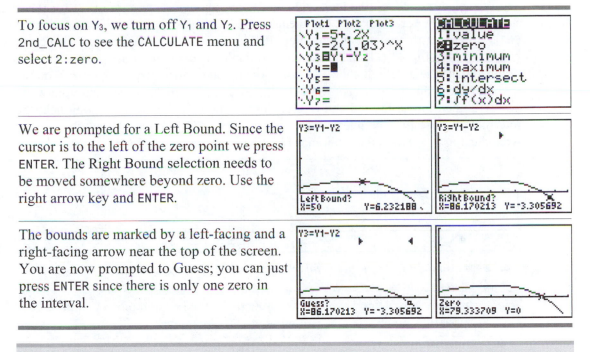

To focus on $Y_3$, we turn off $Y_1$ and $Y_2$. Press 2nd_CALC to see the CALCULATE menu and select 2:zero.

We are prompted for a Left Bound. Since the cursor is to the left of the zero point we press ENTER. The Right Bound selection needs to be moved somewhere beyond zero. Use the right arrow key and ENTER.

The bounds are marked by a left-facing and a right-facing arrow near the top of the screen. You are now prompted to Guess; you can just press ENTER since there is only one zero in the interval.

> *Tip:* The closer the bounds and the better the guess, the faster it finds the zero.

### Finding extrema

The sequence of steps for finding a 3:minimum or 4:maximum is the same as finding a zero above: Left Bound, Right Bound, Guess.

Figure 5.2 shows finding the maximum of the excess food function. Just by looking at the graph on the screen you see that the maximum is between years 20 and 60. This part of the graph is displayed as a horizontal line segment: this is misleading because the actual values of the function are not a constant on this interval. (The resolution of the calculator screen is limited.) After selecting 4:maximum we are asked for the Left Bound, Right Bound, Guess. This time we use an alternative method to set the bounds: enter numeric values instead of using the arrows.

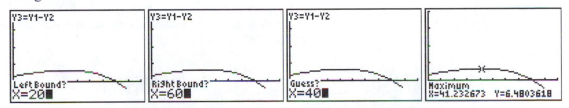

*Figure 5.2 Using 4:maximum to find the maximum of a function within an interval specified by typing.*

> *Tip:* You can normally press ENTER at the Guess prompt, unless you want to speed up the search or you have more than one answer in your interval.

### Finding an intersection

A typical mathematical task is finding where two functions are equal. For Malthus, we found where food supply meets population need by forming a difference function, $Y_3=Y_1-Y_2$, and using 2:zero. But where the graphs of the two functions intersect can be found directly by just using 5:intersect.

We select $Y_1$ and $Y_2$ (the food and population functions) and turn off $Y_3$ for this example. Select 5:intersect from the CALC menu and you are asked to identify the first curve; press ENTER to accept the default choice (identified by the flashing trace cursor). Select the second curve by pressing ENTER. (A choice is available in case there are more than two curves graphed. Change a default choice by using the up and down arrows.) Now it prompts you for a guess; press ENTER to accept the default. (In cases where there are multiple intersections, you will need to use the left and right arrows to move the cursor to a point near the desired intersection.) We find the intersection point is at 79.333709 years. We got the same answer as we did using 2:zero on the difference function — we must have done it right!

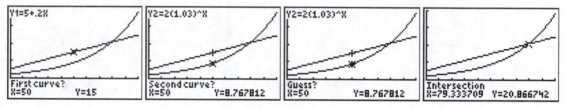

*Figure 5.4 Using* 5:intersect *to find the intersection of two graphs.*

> **Tip:** In cases where you have only two functions graphed and only one intersection appearing in the window, just press ENTER three times.

### Calculus values: CALC 6:dy/dx and 7:∫f(x)dx

These two menu items are fundamental to calculus and are introduced and used extensively in Part II and III.

# The FORMAT and STAT PLOT menus

The 2nd_FORMAT menu allows you to change or embellish the appearance of the basic graph. We use it later in the Appendix to change to polar coordinates, but otherwise the default settings (those on the left) are recommended.

The 2nd_STAT PLOT menu is used to plot statistical data and. If the top row of the Y= menu has Plot1, Plot2, or Plot3 highlighted, you can turn them off by using the navigation arrows to highlight the plot item and pressing ENTER. It is a toggle, like the = sign for displaying functions. If you experience extra graph images or unexplained error screens when you graph, then you may have one of the plots turned on.

*Figure 5.5 The FORMAT menu.*

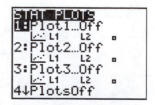

*Figure 5.6 The STAT PLOT menu.*

# CHAPTER SIX

# SOLVING EQUATIONS

## Solving a quadratic equation: MATH Solver

Mathematicians don't use many verbs. We mostly say "this equals that" and shout a few commands like: find, evaluate, simplify, and solve. What does it mean to *solve* an equation? For the equation $3x + 2 = 0$, the solution is $x = -2/3$. The quadratic equation $x^2 - 2x - 3 = 0$ has two solutions, $x = 3$ and $x = -1$. In these examples there is just one variable and we want to know for which value (or values) of the variable the equation is true. If we switch to function notation and let $p(x) = x^2 - 2x - 3$, then solving the quadratic equation is the same as finding the zeros of the function $p(x)$. Thus, in this case and many other cases, the Solver is really not essential; you could *solve* the equation by graphing the function and using the techniques of the last chapter to find the zeros. But the GRAPH and CALC approach are time consuming, so if you want a quick answer, use Solver. Let's find the zeros of the polynomial $p(x)$ given above.

| | | |
|---|---|---|
| Press MATH to see the menu choices. Ø:Solver...is only visible after scrolling, but by pressing the up arrow you can wrap around scroll to it. | MATH NUM CPX PRB<br>1▪▶Frac<br>2:▶Dec<br>3:³<br>4:³√(<br>5:ˣ√<br>6:fMin(<br>7↓fMax( | MATH NUM CPX PRB<br>4↑³√(<br>5:ˣ√<br>6:fMin(<br>7:fMax(<br>8:nDeriv(<br>9:fnInt(<br>Ø▪Solver… |
| In the first stage, you enter the function and press ENTER. If a previous function shows in the first step press CLEAR. | EQUATION SOLVER<br>eqn:0=▪ | EQUATION SOLVER<br>eqn:0=X²-2X-3▪ |
| The screen changes in the second stage, showing the variable at some value (maybe different from what is shown). We accept the default guess and press ALPHA_SOLVE. The solution is shown with a square mark to indicate it has been calculated. | X²-2X-3=0<br> X=79.33370870▪<br> bound={-1ᴇ99,1… | X²-2X-3=0<br>▪X=3▪<br> bound={-1ᴇ99,1…<br>▪left-rt=0 |
| To find the other solution, make a guess closer to -1. The solution looks funny; it must be interpreted by you as -1. If you enter a guess of -0.5, an exact answer of -1 is shown. | X²-2X-3=0<br> X=0▪<br> bound={-1ᴇ99,1…<br> left-rt=0 | X²-2X-3=0<br>▪X=-.999999999▪…<br> bound={-1ᴇ99,1…<br>▪left-rt=0 |

In some cases the calculation never reaches zero exactly; the `left-rt` indicator (left side minus right side) tells you the exactness of the solution. There are some deep mathematical problems concerning rounding errors and exactness because this is a finite decimal place calculator, but in general the results are trustworthy. To restrict the solution search to within some boundaries, reset `bound=` with an upper and a lower bound.

## The no solution message: `ERR:NO SIGN CHNG`

What happens when there is no real solution to an equation? For example, find the zeros of the quadratic function $q(x) = x^2 + 2x + 3$ (this is very similar to the polynomial we solved above). Figure 6.5 shows the surprising message `ERR:NO SIGN CHNG`. This is not really an error, rather the calculator is reporting that it found your expression did not change signs on the interval designated. So what? The way that the `Solver` finds a zero is to find two function values, one positive and one negative. It knows a zero is sandwiched in between and it uses secant lines to hone in on the zero. (If it doesn't start with one positive and one negative value then it still tries to hone in on a value within the tolerance of zero.) If it can't find a sign change (or reach a tolerable zero), then it gives you the message about why it failed.

```
X²+2X+3=0           ERR:NO SIGN CHNG
 X=-.999999999■...  1█Quit
 bound={-1ᴇ99,1...  2:Goto
 left-rt=2
```

*Figure 6.1 Solver finds no real solution.*

## Analyzing investments using Solver

The previous example does not show off the true power of the `Solver`. Commonly, equations have several variables, and the `Solver` can be very handy in these cases. For example, the formula for calculating the growth on a continuously compounded investment (a type of exponential growth) is given by

$$P = P_0 e^{kt}$$

where $P$ is the future worth, $P_0$ is the present value, $k$ is the rate of return, and $t$ is the time (in years) of the investment.

Typically in an investment opportunity, you may be asking any one of four following questions:

1.  What will my investment be worth at some future date? (Find $P$, knowing the other variables.)
2.  What will I need to investment now in order to get a desired amount in the future? (Find $P_0$, knowing the other variables.)
3.  What investment rate do I need in order to have a desired amount in the future? (Find $k$, knowing the other variables.)
4.  How long will it take my investment to be a desired amount in the future? (Find $t$, knowing the other variables.)

Numeric variables must be single capital letters, so we choose $A$ for the initial (or present) amount $P_0$, $B$ for our future amount, $K$ for our investment rate, and $T$ for time.

Press MATH 0:Solver… and clear the previous equation, if any. Enter the new equation as shown in the following sequence of steps. Enter any three of the equation variables and use ALPHA_SOLVE to calculate the value of the remaining variable. You can stay in the solve mode and play "what if."

| | | |
|---|---|---|
| Enter the Solver, the previously solved equation is still there, so arrow up to the equation, and press CLEAR.   Enter your equation, press ENTER. A list of variables and their current values appears. (Values vary.) | EQUATION SOLVER<br>eqn:0=B-Ae^(K*T)<br>■ | B-Ae^(K*T)=0<br>B=3■<br>A=2<br>K=0<br>T=0<br>bound={-1E99,1… |
| Let's answer a type 1 question: What is the future value of $1000 at 6% in 10 years? To find this, enter the values for $A$, $K$, and $T$, then arrow up to $B$ and press ALPHA_SOLVE.<br><br>(Answer: $1822.12) | B-Ae^(K*T)=0<br>B=3■<br>A=1000<br>K=.06<br>T=10<br>bound={-1E99,1… | B-Ae^(K*T)=0<br>•B=1822.118800■…<br>A=1000<br>K=.06<br>T=10<br>bound={-1E99,1…<br>•left-rt=0 |
| To answer a type 2 question, suppose our goal is to have $1500 in 10 years with a 6% rate. Enter 1500 for $B$, arrow down to $A$, and press ALPHA_SOLVE.<br><br>(Answer: $823.22) | B-Ae^(K*T)=0<br>B=1500<br>A=1000■<br>K=.06<br>T=10<br>bound={-1E99,1…<br>left-rt=0 | B-Ae^(K*T)=0<br>B=1500<br>•A=823.2174541■…<br>K=.06<br>T=10<br>bound={-1E99,1…<br>•left-rt=0 |
| For a type 3 question, suppose we are promised $1500 in 10 years on an initial investment of $500. What is our rate of return?<br><br>(Answer: ≈11%) | B-Ae^(K*T)=0<br>B=1500<br>A=500<br>K=.06■<br>T=10<br>bound={-1E99,1…<br>left-rt=0 | B-Ae^(K*T)=0<br>B=1500<br>A=500<br>•K=.1098612288■…<br>T=10<br>bound={-1E99,1…<br>•left-rt=0 |
| Time is the unknown in a type 4 question. This specific answer tells us how long it takes $500 to triple at 6%.<br><br>(Answer: ≈18.3 years) | B-Ae^(K*T)=0<br>B=1500<br>A=500<br>K=.06<br>T=10■<br>bound={-1E99,1…<br>left-rt=0 | B-Ae^(K*T)=0<br>B=1500<br>A=500<br>K=.06<br>•T=18.31020481■…<br>bound={-1E99,1…<br>•left-rt=0 |

*Tip:*    Many menu choices paste expressions to the current cursor location. Pay attention to the cursor so that you don't paste to unexpected places.

# CHAPTER SEVEN

# THE LIMIT CONCEPT

A fundamental difference between precalculus and calculus is the application of the limit. In precalculus, we can define average velocity over a time period of some positive length. For example, if you drive 200 miles in four hours, then you averaged 200/4 = 50 miles per hour. But looking at the car's speedometer, you see the speed at a given moment, the instantaneous velocity. In this chapter we use tables and graphs to investigate the idea of the limit.

## Creating lists

Table 7.1 shows heights of a grapefruit thrown in the air. Find the average velocity over periods of one second. We use STAT, 1:Edit, to access a column display of lists, $L_1$, $L_2$, ..., $L_6$. The time values are in $L_1$ and height values are in $L_2$ as shown in Figure 7.1. It is also possible to enter list values directly from the home screen, for example: {0,1,2,3,4,5,6}→$L_1$.

| time (seconds) | 0 | 1 | 2 | 3 | 4 | 5 | 6 |
|---|---|---|---|---|---|---|---|
| height (feet) | 6 | 90 | 142 | 162 | 150 | 106 | 30 |

*Table 7.1 Grapefruit height, per second.*

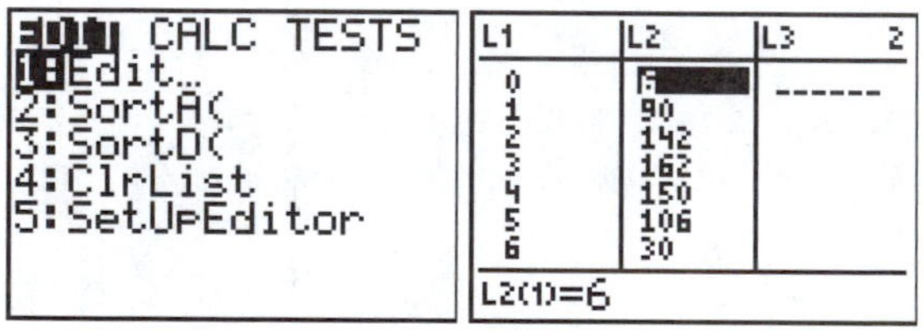

*Figure 7.1 Entering the data into two lists using* STAT, 1:Edit.

*Tip:* In case a list already has values in it, you can clear it by using the up arrow to highlight the list name and pressing CLEAR ENTER.

Now we would like to find the average velocity over the first second of time,

$$(90-6)/(1-0) = 84$$

The TI does this kind of repetitive calculation and displays the whole list at once. We use the command ΔList(. This is in the catalog as shown in Figure 7.2 (Or it can be pasted from the 2nd_LIST, OPS menu.) This command finds the successive differences of a list and then the lists are used to divide term by term. In this case all the time intervals are of length one so division is unnecessary. The full answer list is seen by horizontal scrolling.

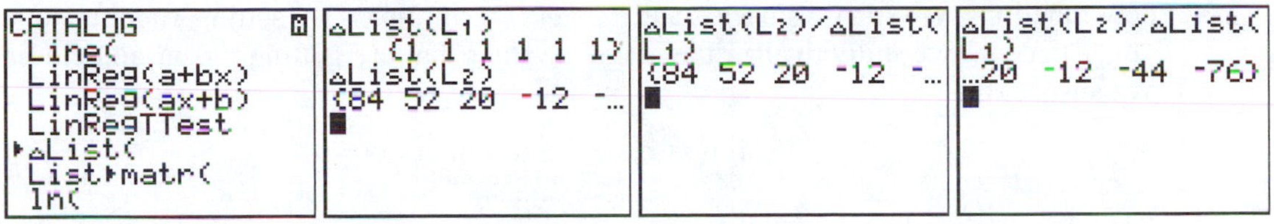

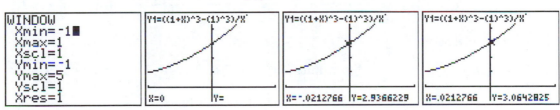

*Figure 7.2 Using △List to find the average velocity from two data lists.*

## What does the lim notation mean?

The key to the calculation of an instantaneous velocity is to let the time period become closer and closer to zero. Notice that the time period is in the denominator of the average velocity, so letting it reach zero would mean dividing by zero — a definite error. Let's see how the limit notation, $\lim_{x \to 0} f(x) = L$, works with two function examples:

$$f(x) = x^2 + 3 \text{ and } g(x) = \frac{(1+x)^3 - (1)^3}{x}.$$

The first function, $f$, is defined for all $x$, and at $x = 0$, $f(0)=3$. It is no mystery to write

$$\lim_{x \to 0} f(x) = 3.$$

In Figure 7.3, we graph the second function as Y₁ and in the second frame use TRACE to see that it is not defined at $x = 0$. But looking at the $y$-values of $g(x)$ on each side of zero we see one is 2.9366229 and the other 3.0642825; it doesn't take a rocket scientist to figure out that the values are getting close to 3. This is what we mean when we write

$$\lim_{x \to 0} g(x) = 3.$$

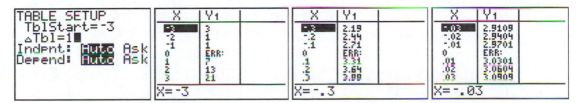

*Figure 7.3 The graphical meaning of the limit notation. The function graph looks like f(0)=3..*

Now let's use the table approach to see the same thing. We get closer and closer by successively using △Tbl = 1, 0.1, 0.01, etc. See Figure 7.4.

*Figure 7.4 Using TABLE and TBL SET to find function values near 0.*

### An unreliable table

Let's look at two examples, in one, our limit guess is wrong. In Figure 7.5, we define our two new functions, $Y_1 = f(x) = \sin(x)/x$, and $Y_2 = g(x) = \sin(2\pi/x)$. By setting Indpnt: to Ask in the TBLSET menu, we individualize the set of $x$-values that are getting closer and closer to zero. We suspect that

$$\lim_{x \to 0} f(x) = 1 \quad \text{and} \quad \lim_{x \to 0} g(x) = 0 .$$

Enter a value, such as 0.003, that is not the reciprocal of an integer, and you see the function values of $g(x)$ are not going to zero. The limit of $g(x)$ as $x$ approaches 0 does not exist. The $g(x)$ function is similar to the one used for the 'inaccurate graph' from Chapter 4. Its graph hints that it has no limit.

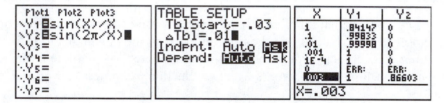

*Figure 7.5 The graphical meaning of the limit notation.*

## Speeding ticket: the Math Police let you off with a warning

A warning: The graphic and table evidence may be very strong in indicating what a limit value should be, but don't make a speedy decision, this can lead to the wrong conclusion. Only by using careful mathematical analysis can you really prove that the limit exists. For example, we rewrite

$$g(x) = \frac{(1+x)^3 - (1)^3}{x} = \frac{(1 + 3x + 3x^2 + x^3) - 1}{x} = 3 + 3x + x^2$$

and see that

$$\lim_{x \to 0} g(x) = \lim_{x \to 0} (3 + 3x + x^2) = 3 .$$

Other techniques are necessary for trigonometric functions.

# PART II  DIFFERENTIAL CALCULUS

# CHAPTER EIGHT

# FINDING THE DERIVATIVE AT A POINT

The derivative at a point is the instantaneous rate of change that we mentioned in the previous chapter. The average rate of change in miles per hour for a trip in your car is calculated by dividing the distance by the time. The instantaneous rate of change is what is shown continuously by the speedometer as you drive. In this chapter we look at a graphical interpretation of the derivative at a point and learn two ways to use the TI to calculate this value.

## Slope line as the derivative at a point

In the last chapter we saw that we could set a center point and zoom in for a microscopic view of a graph. By continued zooming, the graph began to appear linear. If we zoom in until the graph appears linear, then we are essentially seeing the tangent line to the graph. The slope of this tangent line to the graph is the derivative at the center point. You use the DRAW menu to draw a tangent line to a graph at a given point.

In Figure 8.1, we start with $f(x) = x^2$ and draw a tangent line on the graph at $x = 1$. We graph using a ZDecimal window, then use 2nd_DRAW 5:Tangent. To select $x = 1$, arrow across or enter the value. Press ENTER. Notice that the equation of the tangent line is given at the bottom of the screen. The slope of the tangent line (or the derivative of the function) at the point $x = 1$ is 2.

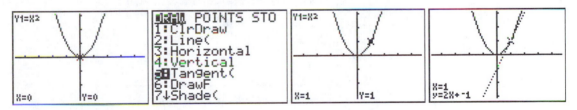

*Figure 8.1 Using the DRAW menu to find a tangent line.*

*Tip:*  If you want a second tangent line, the first one remains on the screen unless you use 1:ClrDraw to start with a fresh graph.

# The numerical derivative at a point

Tangent lines are like training wheels on a bicycle, eventually they become unnecessary. Next we see how to numerically find the derivative at a point. In Leibniz notation, $dy/dx$ is the symbol for the derivative. This selection is available in the CALC menu. Figure 8.2 shows finding the derivative of the sine function at $x = 1$. This derivative value, about one-half, is less than the derivative of $f(x) = x^2$ at 1; in comparison, the function values of the sine function are increasing more slowly near this point.

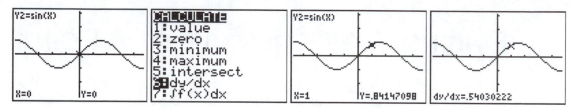

*Figure 8.2 Using* CALC  6:dy/dx *to find the derivative at a point on a graph.*

### The numerical derivative at a point, without a graph

We can skip the graph and find the derivative of the function (Y₂) from the home screen by using a MATH menu command,

$$\text{nDeriv}(Y_2, X, 1)$$

This command doesn't look like any standard calculus notation, so let's translate it into calculus: the name, as you may have guessed, is an abbreviation of "numerical Derivative." The first entry is always the function, the second entry is the independent variable of the function, and the third entry is the point at which to evaluate. A function to be used in nDeriv need not be stored in the Y= editor and it need not have $x$ as the variable. For example, you can enter nDeriv(T², T, 1) and get a value of 2 (as found for Y₁=X²).

Proceed as follows. First 2nd_QUIT the graph screen and CLEAR the home screen. Now press MATH and arrow down to nDeriv(. Press ENTER to paste it on the screen. Then use the well-known sequence VAR Y-VAR 1:Function… to paste Y₂ in as the function. Finish the entry and press ENTER

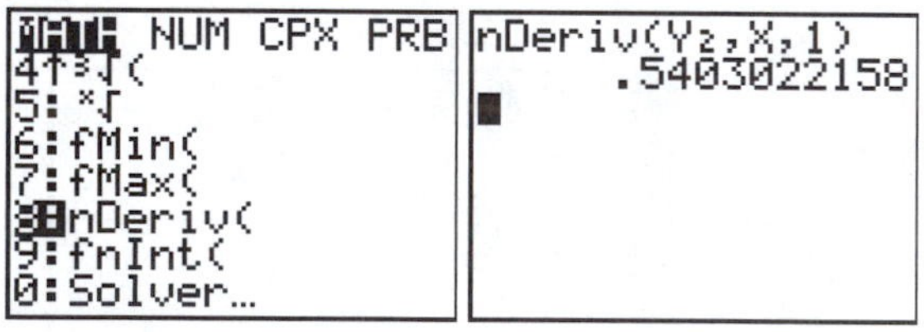

```
MATH NUM CPX PRB      nDeriv(Y₂,X,1)
4↑³√(                      .5403022158
5: ×√                 ▮
6:fMin(
7:fMax(
8:nDeriv(
9:fnInt(
0:Solver…
```

*Figure 8.3 Using  MATH 8:nDeriv to find the derivative of a function at a point.*

# How does the TI calculator find the numerical derivative?

You know from your text that the derivative of $f$ at $a$ is defined as

$$\lim_{h \to 0} \frac{f(a+h) - f(a)}{h}, \text{ if the limit exists.}$$

The calculator finds a single value to report as the derivative at point *a* using the related formula

$$\text{nDeriv}(f(\text{X}),\text{X},a) = \frac{f(a+h)-f(a-h)}{2h}, h = 0.001$$

## Warning: Approximations can give false results

It is important to realize that nDeriv is an *approximation*. This short cut can get us into deep trouble with certain points of some functions. In the first frame of Figure 8.4, we see the function $f(x) = 1/x$ is not defined at zero and thus can not have a derivative there. In the second frame the function $g(x) = x^{\wedge}(1/3)$ is defined at zero but the tangent line is vertical and the derivative at that point does not exist. But using MATH 8:nDeriv( erroneously gives a values of 1 million and 1 hundred for the two numeric derivatives at zero.

The lesson here is that blindly using nDeriv is dangerous. Using a graphical approach is recommended when the function is not well known to you. If the function is undefined at a point then 2nd_CALC,8:dy/dx does not give a numeric derivative there. However, in cases where the graph is defined but the tangent line is a vertical line, then an erroneous value is calculated.

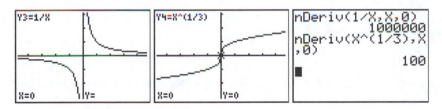

Figure 8.4 *The potential problems with* nDeriv *when the derivative is undefined: false values.*

> **Tip:**  Be careful when finding the derivative at a point: be sure the function is defined at the point, and the tangent line is not vertical.

## Improving the accuracy of an approximation (Optional)

The default value $h = 0.001$ can be changed to increase the accuracy of an approximation. However, as we try to increase the accuracy by making *h* extremely small, we run afoul of the calculator's computational restrictions and our accuracy declines. We see this in Figure 8.5, where we use a fourth, optional, entry of nDeriv to set *h*. Of the five values, the second value ($h = 0.00001$) is closest to the true value of the derivative of $2^{\wedge}x$ at $x = 0$. That true value, to ten decimal places, is 0.6931471806.

```
nDeriv(2^X,X,0)    nDeriv(2^X,X,0,E
       .6931472361  -10)
nDeriv(2^X,X,0,E            .69305
-5)                nDeriv(2^X,X,0,E
       .693147179   -15)
                            0
```

Figure 8.5 *Resetting a default to increase accuracy and going too far.*

> **Tip:**  Use 2nd_EE (above the comma) to enter small values. Use negation, not subtraction.

# CHAPTER NINE

# THE DERIVATIVE AS A FUNCTION

Once we know the value of the derivative at a whole set of points, then we define a new function called the derivative of $f$:

$$f'(x) = \lim_{h \to 0} \frac{f(x+h) - f(x)}{h}, \text{ if the limit exists.}$$

The derivative at a point was defined in the last chapter and we showed that the MATH menu command

$$\texttt{nDeriv(T²,T,1)}$$

gave a numerical approximation of the derivative of $f(T) = T^2$ at the point $T = 1$. We now find this new derivative of $f$ at any point $x$ by defining:

$$f'(x) = \texttt{nDeriv(T²,T,X)}$$

In this chapter we show how to graph this function and better understand the relationship between a function and its derivative.

## Viewing the graph of a derivative function

We know T is the dummy variable in the above function definition, so why not call it $x$? We can, at the risk of some confusion. The first time you see $\texttt{nDeriv(X²,X,X)}$, you might think there is an extra $x$. We started in the last chapter with the function $T^2$ and variable $T$ to help make it clear here that not all these $x$'s are being used in the same way. In the first two entries of $\texttt{nDeriv(X²,X,X)}$, the $x$ is a dummy variable, while in the third entry, it is the independent variable. This is a confusing aspect of the TI notation that we live with.

### Matching a function to the graph of its derivative

We consider three classic functions:

$$f(x) = x^2, \; g(x) = \sin(x) \text{ and } h(x) = x^{1/3}.$$

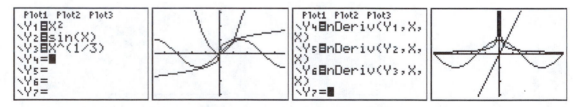

*Figure 9.1 Matching: three classic functions to three derivative functions.*

The three functions were stored in $Y_1$, $Y_2$, and $Y_3$ and graphed using a ZDecimal window setting. Now we graph the derivative function for each of the three functions and display them in Figure 9.1. Can you match the curves to the derivative functions?

We could, of course, graph them one by one, but using a little thought we can identify them just by looking at the features of the graph. Consider the parabolic function $f$: it is decreasing until it reaches zero, then it is increasing. Since the derivative gives the instantaneous rate of change, this means that the derivative values are all negative to the left of the origin and are all positive to the right of the origin. Of the three options, this describes the line. (You may also know the power rule of derivatives which says that the derivative of a quadratic function is a linear function). Now consider the sine function: it oscillates between decreasing and increasing, so the derivative should oscillate between negative and positive. There is only one function that does this and it looks like a cosine function (another rule you may know). The remaining derivative function is always positive and has a spike at zero; this fits the slope patterns of the Y₃ graph. As a reminder, the derivative function of $h(x)$ is undefined at zero, but the calculator erroneously gives it a value there.

## Graphing a function and its derivative in the same window

It is instructive to graph a function and its derivative in the same window, but you may want a means of distinguishing which is which. This is accomplished by using a style feature in the Y= edit menu. Arrow left to the slash sign on the left of the Y=. You change it to a bold line by pressing ENTER. Unfortunately, to change back to the regular style, you must press ENTER six times to cycle through the entire set of styles. The derivative is bold.

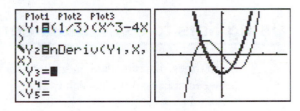

Figure 9.2 Two styles to distinguish functions.

Functions and their derivatives do not always fit well in the same window. Consider a slight modification to the above example and add 100 to the Y₁ function, shown in Figure 9.3. The function graph will not appear in the ZDecimal window, but the derivative function will be identical to the derivative function shown in Figure 9.2.

There is a lesson here. Showing the graph and its function in the same window is a parlor trick and must be carefully designed to work. Further, it should be realized that a function serving as a model and its derivative will use different interpretations of the $y$-axis. For example, when modeling motion, the distance function might be in the feet and the derivative function would be in feet per second.

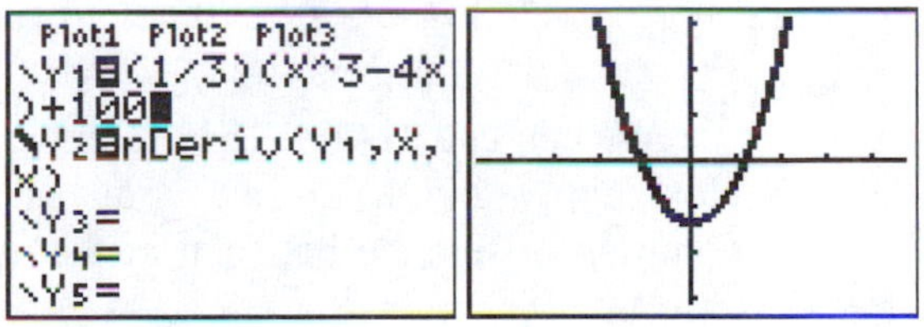

Figure 9.3 The function graph and derivative graph rarely fit in the same window.

# The function that is its own derivative

Could some function be its own derivative? To save time in guessing, we will try an exponential function. In Figure 9.4, first change the function Y₁ to the doubling function $2^x$ and graph it along with its derivative. Since we are trying to match function graphs we switch the derivative graph style from bold to path, this style draws the graph path with a bubble head and it is easy to see when it is writing over another function. The graphs are similar. Next change Y₁ to be a tripling function $3^x$ and graph again. This function's graph is very close to the graph of its derivative function. The exact exponential function we are looking for has a base between 2 and 3. In calculus, we find this amazing number is 2.718..., an irrational number denoted by *e*. Change Y₁ to be e^(x) and graph again. Perfect fit!

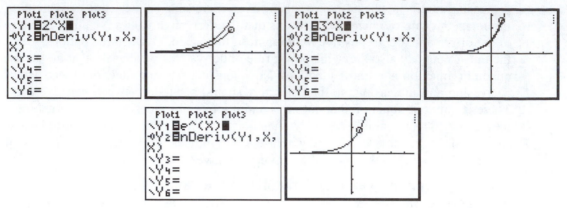

*Figure 9.4 Looking for a function that is its own derivative. Graphing y=e^x, using a bubble path style to distinguish two functions whose graphs are similar.*

## Using lists to estimate a derivative function

Sometimes discrete data is gathered and we use it to construct the average rate of change between successive data values. For example, consider the following table showing distance traveled by a car.

| Time (seconds) | 0 | 3 | 6 | 9 | 12 |
|---|---|---|---|---|---|
| Distance (meters) | 0 | 60 | 159 | 288 | 441 |

*Table 9.1 Distance traveled by a car.*

We can use STAT 1:Edit to enter the times and distances as list in L₁ and L₂, respectively. Next apply 2nd_LIST OPS 7:ΔList to the list and then divide by ΔList(L₁). (Since the common difference between measurement times is equal, you could just divide by 3 in this case.)

Store these difference quotient values in L₃, as shown in Figure 9.5. However 20 is first in the L₃ list, but it is our estimate for time 3 seconds, our second time entry. Thus our interpretation needs to be read by shifting down L₃ values by one place. In this example, we interpret the values shown in L₃ as the velocities.

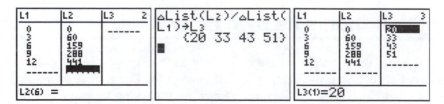

*Figure 9.5 Making a list of numerical derivatives.*

# CHAPTER TEN

# THE SECOND DERIVATIVE: THE DERIVATIVE OF THE DERIVATIVE

The derivative function for the function $Y_1$ was created by setting

$$Y_2 = nDeriv(Y_1, X, X)$$

Since $Y_2$ is a function itself, there is no stopping us from writing the *derivative of the derivative* as a function:

$$Y_3 = nDeriv(Y_2, X, X)$$

Using traditional calculus notation, we would say

$$f'(x) = nDeriv(Y_1, X, X) \text{ and } f''(x) = nDeriv(Y_2, X, X)$$

In this chapter we see how the second derivative can tell us about the concavity of the graph of the original function.

## How to define and graph f(x), f'(x), and f''(x)

Let's start with a known result. If $f(x) = x^2$, then by the power rule $f'(x) = 2x$ and $f''(x) = 2$. In Figure 10.1, we can see the graphs of these three functions as a parabola, a positive sloped line and a horizontal line.

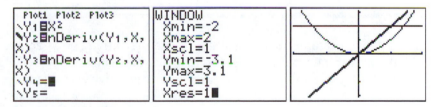

*Figure 10.1 A graph of $y = x^2$ with its first and second derivatives.*

> **Tip:** Graphs using nDeriv in their definition require massive calculations that slow the graph process considerably. For slower TI models, setting Xres = 3 increases the speed by graphing only one-third of the usual *x* values.

We previously commented that, in general, trying to graph a function and its derivative in the same window is not practical. We see this again in the next example as we analyze the logistic function and its derivatives.

# Looking at the concavity of the logistic curve

In most growth situations, a logistic growth model makes more sense that a pure exponential model. A new software company may keep doubling employees, but the growth has to slow down or else the number of employees would eventually be out of this world, literally. Recall how in Part I we found that double folding a paper 42 times would reach the moon.

| | | |
|---|---|---|
| Enter Y1 =1000/(1+9e^(-.05x)) as the logistic function giving the number of employees in a company at time $x$. ZStandard gives as inadequate window to view the graphs of all three functions. | ```Plot1 Plot2 Plot3 \Y1 ⊟1000/(1+9e^( -.05X))∎ \Y2⊟nDeriv(Y1,X, X) \Y3⊟nDeriv(Y2,X, X) \Y4=``` | |
| We set the window so that the domain includes all the important behavior of the logistic function graph. Y1 has a range of values that makes seeing Y2 and Y3 impossible in this window. | ```WINDOW Xmin=-10∎ Xmax=200 Xscl=10 Ymin=0 Ymax=1200 Yscl=100 Xres=1``` | |
| Reset the window for Y2. The function Y1 is monotonically increasing on the interval shown, so the derivative function Y2 must be positive. Notice that Y2 peaks and decreases to zero. The peak is at the point of the fastest growth of Y1. | ```WINDOW Xmin=-10 Xmax=200 Xscl=10 Ymin=0 Ymax=15 Yscl=5∎ Xres=1``` | |
| Reset again to see Y3. The second derivative shows the rate of change of the rate of change. The value of Y3 is zero at the peak of Y2. | ```WINDOW Xmin=-10 Xmax=200 Xscl=10 Ymin=-.5 Ymax=.5 Yscl=.1 Xres=1∎``` | |

In the above example we see that the second derivative (Y3) is zero at about 40. This is because the growth rate (Y2) peaks at about 40 and begins to slow. On the graph of Y1, this is the point of fastest growth. It is a point where concavity is changing from positive to negative. We call this a point of inflection on the graph. Since this is an important point on Y1, we identify it more exactly.

Our way of identifying this point is to find the zero of the second derivative. The computation is shown in Figure 10.2, but be warned that it takes many seconds on older model calculators. Finding the zero of the second derivative is much faster when an actual algebraic formula is known.

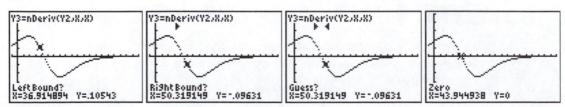

*Figure 10.2 Using* CALC, 2:Zero *to find the zero of the second derivative function from the graph.*

# Creating a second derivative table

We find the derivative of the velocity function in the same way as we found the derivative of the distance function. The second derivative of the distance function is the acceleration. In the last chapter, recall we the calculated a car's velocity at four different times. We return to these lists using STAT 1:Edit to begin where we left off. Move to the heading L4 and press ENTER to begin entering our formula Y4=ΔList(L3)/3. Press enter to calculate L4 as shown in Figure 10.3. Note that since all the time intervals are equal, dividing by 3 simplifies the calculations.

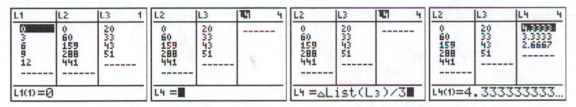

*Figure 10.3 A discrete list of distance, velocity and acceleration over time*

In L3 we are faced with the fact that the corresponding values are one off from L1, now the values of L4 are two off from L1 and there are only three values in calculated. To make these tables less confusing, we insert a known value for time at zero: the velocity and acceleration are both zero there. In the first frame of Figure 10.4, we arrow to position L3(1) and press 2nd_INS to insert a zero. Now all the velocities are in line with the times in L1. We must now recalculate the acceleration column L4 and use 2nd_INS on position L4(1) to arrive at a more readable table.

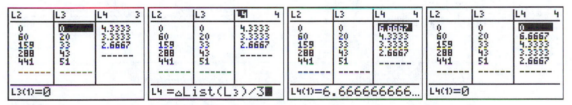

*Figure 10.4 Insert 0 in L3, recalculate, and insert 0 in L4 to get a more readable table.*

## What does acceleration mean?

In the above example, for the time measured, the car was

- going forward: this is the distance list, L2.
- going with an increasing velocity: this is the list L3.
- going with a decreasing acceleration: this is the new list, L4.

Although the velocity is increasing, it increases by less and less each three second interval. So the acceleration, the rate of change of velocity, is decreasing. Let's face reality: a car in low gear can peel out of a parking lot and make a squeal, but at higher speeds there is less acceleration making squealing nearly impossible.

# CHAPTER ELEVEN

# THE RULES OF DIFFERENTIATION

Using the definition of the derivative to find a derivative function is cumbersome; fortunately there are shortcuts to finding derivative functions. We use the calculator to see examples of the main three rules. These rules must be proved analytically, but a graphical verification gives us added confidence in them. Also, we can show graphically that some common guesses for the rules are wrong.

## The Product Rule

In Figure 11.1, we take two common functions and graph the product function and the derivative of the product function. We see from the graph that the product function has a local extrema at $x = -2$, which is heralded by the derivative function being zero. (Y=-4.512E-8 is essentially zero).

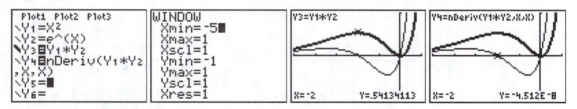

Figure 11.1 A product function and the derivative of the product function.

The product rule cannot be proved by graphing, but in Figure 11.2 we enter the correct rule in $Y_5$ and see that it matches $Y_4$ in Figure 11.1. Further, let's suppose a guess for the product rule is that that derivative of the product is the product of the derivatives. This is not such a silly guess since the derivative of a sum is the sum of the derivatives; but it is wrong. The product of the derivative functions, $Y_6$, does not have the correct graph nor does it have a zero where it should, at $x = -2$, so it is not the derivative of the product function.

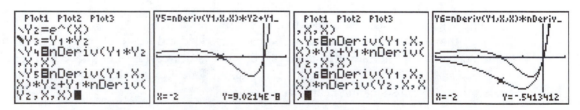

Figure 11.2 The product rule graph and graphical proof that $(fg)' \neq f'g'$.

***Tip:*** To see if two graphs are equal, graph them in the same window, then in TRACE mode use the up/down arrow keys to compare two graphs at a point.

> *Tip:*    As an alternate method to see if two graphs are equal, set the graph style of the second function to be Path (a bubble icon). The graph of the second function traces the first with a visible bubble.

## The Quotient Rule

Let's check the quotient rule in the same way. This means we can recycle the functions, make a minor change to the window setting, and get the graph shown in Figure 11.3. Notice the two zeros of the derivative function are at the local maximum and minimum of the quotient function.

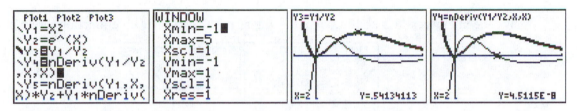

*Figure 11.3 A quotient function and its derivative.*

We can again make a feasible, but incorrect, guess that the derivative of the quotient is the quotient of the derivatives. We see our folly in Figure 11.4 since the quotient of the derivatives is not zero as $x = 2$. Now enter the correct formula in $Y_6$ and see that it traces over $Y_4$.

*Figure 11.4 The incorrect quotient function, $Y_5$, does not trace $Y_4$, but the correct formula, $Y_6$, does trace the derivative of the quotient function.*

## The Chain Rule

Thinking of a function as a composite function and using the chain rule often simplifies finding the derivative of a function. Consider $y = (x^2+1)^{100}$. A straightforward but impractical approach is to expand the expression, write it as a polynomial of degree 200, and then differentiate term by term. Instead, we apply the chain rule and find the derivative quickly and easily:

$$y' = 100(x^2+1)^{99}(2x) = 200x(x^2+1)^{99}$$

Figure 11.5 shows the graphs of $y$ and $y'$ (notice the scale).

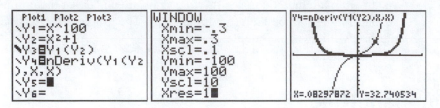

*Figure 11.5 A composite function and its derivative.*

Some care should be exercised in entering the function for the chain rule formula. The derivative of the outside function is evaluated in terms of the inside function, so the outside function is evaluated at $Y_2$. Hence for this calculator we write

$$\texttt{nDeriv(Y}_1\texttt{,X,Y}_2\texttt{)*nDeriv(Y}_2\texttt{,X,X)}$$

We see in Figures 11.5 and 11.6 that the graphs of the two derivative functions are the same.

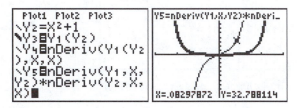

*Figure 11.6 The correct formula, $Y_5$, traces the derivative of the composite function.*

## The derivative of the tangent function

Recall that the window used for graphing trigonometric functions is often crucial. In Figure 11.7 we first use the ZOOM ZStandard window, then the ZOOM ZTrig window, to view $Y_1$ = tan(x). This should remind you that $x$ values are sampled evenly across the window and the function values at these points are then connected to form the graph. It is clear from the graph of the tangent function that it is an increasing function within an interval such as $-\pi/2 < x < \pi/2$ and that it is undefined at multiples of $\pi/2$. Thus we expect the derivative to be positive between $-\pi/2 < x < \pi/2$ and undefined at multiples of $\pi/2$.

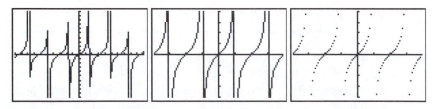

*Figure 11.7 Graph of $y = \tan(x)$ with different windows: ZStandard, ZTrig, and Dot-Style.*

To practice applying the quotient rule, we find the derivative of the tangent function from the quotient definition, $\tan(x) = \sin(x)/\cos(x)$. Set $y = \tan(x)$ and use the quotient rule to derive $y' = 1/\cos^2(x)$. This is always positive and is undefined at multiples of $\pi/2$ (where the cosine is zero).

*Tip:*    The conventional mathematical way of writing a power of a trigonometric function, such as $\cos^2(x)$, gives a syntax error. Use $\cos(x)^2$ or, better yet, $(\cos(x))^2$.

In Figure 11.8, the numeric derivative Y₂ is graphed. This creates a surprise of double vertical lines between the defined intervals. What is going on? If you trace to a value like $\pi/2$ in the second frame, then you see that the numeric derivative has been calculated incorrectly as a large negative number. Enter Y₃ = $1/(\cos(x))^2$ and graph it; it looks perfect.

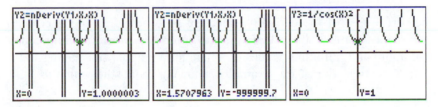

*Figure 11.8 Graph of the numeric derivative and graph of the algebraic derivative of tan(x).*

The moral of this graphing is that we must be constantly vigilant in believing the numeric derivative function at points where the function is undefined. A cosmetic remedy to this problem is to make the sampling avoid the bad places. In Figure 11.9, we raise the Xres setting as shown and it won't evaluate at multiples of $\pi/2$.

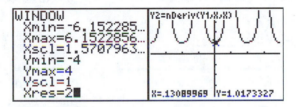

*Figure 11.9 Reset Xres to avoid undefined values.*

## Notes on Xres

The Xres setting can be used on older TI models where calculations and/or graphing are unbearably slow. On all models it has its use as a means of improving the graph appearance in cases where there are discontinuities. An Xres setting of 2, as in Figure 11.9, skips every other $x$ value that the calculator normally uses in its sample of $x$ values between Xmax and Xmin. These 95 samples are formed by adding multiples of (Xmax−Xmin)/94 to Xmin. The only ZOOM setting that changes Xres is ZStandard; it resets to Xres = 1.

# CHAPTER TWELVE

# OPTIMIZATION

One of the powerful uses of the derivative function is to help find the maximum or minimum of a function. But we must confess that calculators and computers with graphing capabilities can, in most cases, find maximum and minimum values of a function without your having to know anything about calculus. In this section there are examples showing both the calculus and the non-calculus approaches.

## The ladder problem

Typically, optimization problems arise from real-world applications. The 'ladder' problem is to determine the longest ladder that can be carried horizontally around a corner that joins two hallways. We assume that the hallways are different widths: the narrower one is 4 feet and the wider one is 8 feet wide. Figure 12.1 shows the position where the ladder could get stuck: it touches both walls and the corner. For the ladder to make this corner, it needs to fit in the hallway for every angle $\theta$, even this tightest one. This leads to minimizing the equation

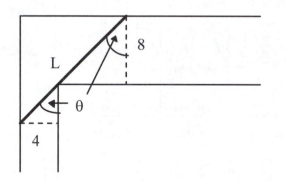

$$L = f(\theta) = 4/\sin\theta + 8/\cos\theta.$$

*Figure 12.1 A ladder carried horizontally around a corner.*

A ladder of that minimal length fits all angles, and is the longest possible ladder that works. Finding such an equation is the hard part. This equation was derived by thinking of the ladder as being divided at the corner point and using the triangle trigonometry definitions for the sine and cosine.

To find the minimum value of $L$, follow these steps:
- be sure that the MODE setting is Radians
- enter the function $L$ in Y₁ with $x$ as the variable
- set the window.

You might be tempted to use ZTrig but, like most models, there is a more restricted domain in this example. The angle $\theta$ is greater than 0 but less than $\pi/2$. To set the window $y$-values, we are generous and say that the minimum ladder is under 50 feet long.

## Find the minimum by trial-and error tracing

Use TRACE to approximate the lowest point on the graph, as shown in Figure 12.2.

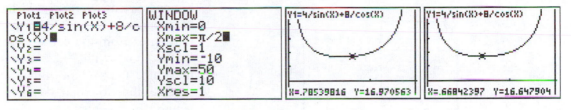

*Figure 12.2 A minimum found by using* TRACE *and the trial-and-error method.*

## Using the CALC feature to find the minimum

Use CALC 3:minimum avoid trial-and-error in finding $y \approx 16.65$. You must provide a Left Bound, a Right Bound, and a Guess. The $x$ values must be within the viewing window. These values can be entered directly or by using the arrows and pressing ENTER.

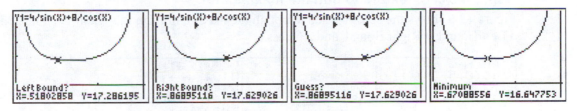

*Figure 12.3 A minimum found by using* CALC 4:minimum.

## Using the graph of the derivative to find the minimum

Calculus tells us that the local maximums and minimums occur at points where the first derivative is zero or undefined. Using styles to distinguish between the curves, we see the function (bold) and its derivative in Figure 12.4. The second derivative is not visible except for a few dots at the top of the screen.

To find the minimum, trace to the closest point to the zero of the first derivative and use the up arrow to read the function value. Use the up arrow again and the value of the second derivative is shown as $y \approx 50$; since this is a positive value, the point is a minimum by the second derivative test.

After all of that, you may wonder why we bother to use the derivatives. One reason is that, in many cases, we can use derivatives to find an exact answer. Such 'closed form' solutions can be important for using in other calculations or for insuring unlimited accuracy.

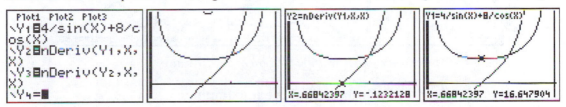

*Figure 12.4 A minimum found by graphing the function and its first and second derivatives.*

# Box with lid problem

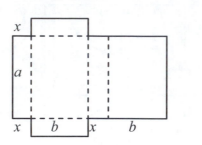

Suppose we have an 8.5 x 11 inch sheet of paper and want to cut squares and rectangles from the corners to create a folded box with lid. We want to maximize the volume. First we find a volume function that depends on the length $x$ of the square's side:

$$V = (8.5 - 2x)(11/2 - x)x$$

*Figure 12.5  A diagram of cutting corners to create a box with lid. Fold on the dotted lines.*

Now we show three different ways to solve this problem.

### Using the CALC feature to find the maximum

Figure 12.6 shows the screens used to obtain the maximum from the graph. You must provide a Left Bound, a Right Bound, and a Guess.

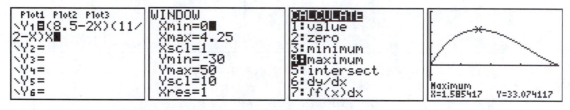

*Figure 12.6 Finding a maximum from the graph. (Bound selection not shown.)*

### Using the zero of the derivative

The zero of the derivative function gives us the size of cut needed. In the last frame of Figure 12.7, we see that the second derivative is negative; this confirms that we are finding a maximum rather than a minimum, which is also obvious from the graph of the original function.

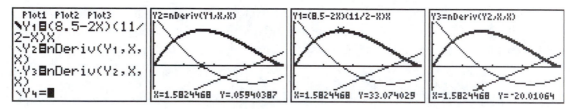

*Figure 12.7 A maximum found by using the zero of the derivative.*

### Using Solver and nDeriv to find the maximum

We can improve on the accuracy of the above answer by either using 2nd_CALC, 2:zero or by Solver to find the zero of the derivative. We use Solver. Recall that you first enter the equation, and press ENTER. Then put the cursor on the variable you wish to know and press APLHA_SOLVE. The calculated answer is shown with a small black box beside it. This answer is stored in the variable X so we can use it in other calculations. In Figure 12.8, we find the maximum volume of the volume function $Y_1$ by using Solver to find the zero of derivative function $Y_2$. When working numerically, without a graph of the original function to show whether it is a maximum or minimum, it is prudent to check the second derivative $Y_3$ to be sure. It is negative, so we have a maximum.

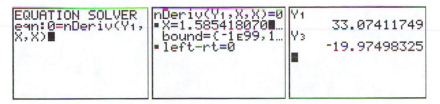

*Figure 12.8 A maximum found by using* So1ver *to give the zero of the derivative and using that stored X to evaluate the function and second derivative.*

## Using the second derivative to find concavity

An important function in statistics is the *standard normal distribution* function. Its graph is bell-shaped and has the definition:

$$f(x) = \frac{1}{\sqrt{2\pi}} e^{-x^2/2}$$

For simplicity we define a function in the same family but one that does not have the coefficient: $1/\sqrt{2\pi}$. The equation and (bold) graph are shown in Figure 12.9. The graphs of the first and second derivatives have been included.

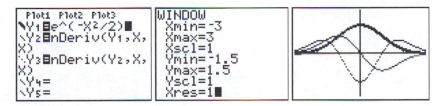

*Figure 12.9 A bell-shaped curve and its derivatives.*

The concavity of the normal curve is revealed by finding the zero of the second derivative. The zero with positive $x$ in this example gives a point where the second derivative changes from negative to positive. That is where the function changes from being concave down to concave up. The zero with negative $x$ gives a point where the function changes from being concave up to concave down. An important characteristic of this bell-shaped curve and its multiples such as the standard normal distribution is that they have points of inflection at $x = \pm 1$.

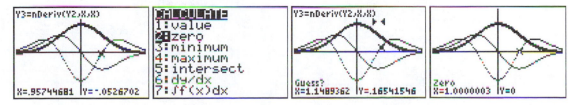

*Figure 12.10 The point of inflection for this bell-shaped curve is at $x = 1$.*

**Notes:**

# PART III INTEGRAL CALCULUS

# CHAPTER THIRTEEN

# LEFT-AND RIGHT-HAND SUMS

The fundamental activity of integral calculus is adding. In the discrete case, we sum a set of values. In the continuous case, we use the integral to sum over an interval. In this chapter we restrict our attention to finite discrete sums. These sums are approximations to the value of a definite integral; we make that connection in Chapter 16, Riemann sums. This chapter is like building a house with toothpicks, unless you are interested in detail, you may want to skip it and use the program in Chapter 16 for left and right sum calculations.

## Distance from the sum of the velocity data

If you drive 50 miles per hour for 3 hours, then you traveled 50 + 50 + 50 = 150 miles. If you drive 20 mph for two hours, then 30 mph for two hours, and finally 40 mph for two hours, then in the six hours you traveled 20(2)+ 30(2) + 40(2) = 180 miles. We rarely travel a constant speed. Table 13.1 shows velocity readings at 6 different times. We do not know if we traveled mostly at 20 ft/sec or 30 ft/sec for the first two seconds. If we assume the velocity is constantly increasing, then these two numbers give us lower and upper bounds for the first two seconds. Repeat this with the last five velocities to find a lower and upper bound on the distance traveled in ten seconds.

| Time (sec) | 0 | 2 | 4 | 6 | 8 | 10 |
|---|---|---|---|---|---|---|
| Velocity (ft/sec) | 20 | 30 | 38 | 44 | 48 | 50 |

*Table 13.1 Velocity of a car, every two seconds.*

### Creating and summing lists from the home screen

The TI has six list variables, named `L1` to `L6`. Lists can be created from the home screen using the set symbols { and } then stored (`STO>`) in one of the list variable names by pressing `2nd_1` to `2nd_6`.After entering the first data list and storing it in `L1` (see Figure 13.1), The sum command is in `2nd_LIST, MATH`. Repeat for the upper sum.

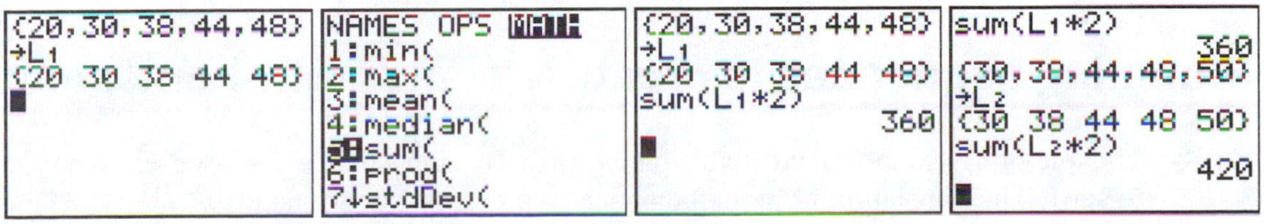


*Figure 13.1 Finding lower and upper bounds, 360 and 420, on distance traveled using lists.*

> *Tip:*    This is a good place to use 2nd_ENTRY to edit the first list and create the second list.

### Creating and summing lists from STAT EDIT

There is an alternate way to enter lists, which we used in Chapters 9 and 10. It is particularly handy when the list is long and when you need to edit the data. In Figure 13.2, we repeat the above bounding computation. Press STAT 1:Edit and notice that data entered as $L_1$ and $L_2$ on the home screen appears in the first two columns, they are the same variable whether edited on the home screen or in STAT EDIT. Arrow down to the bottom of the list and press ENTER, this prompts for an entry. Enter the sum formula just as before using 2nd_LIST, MATH. Repeat for the upper sum.

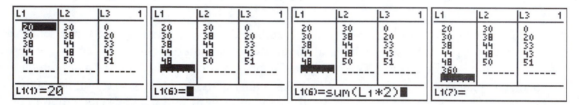

*Figure 13.2 Using STAT EDIT to find a lower bound.*

## Using a sequence to create a list of function values

The sequence command is a very handy feature to make a list of values using an expression with an index that takes on values from a starting point to a stopping point, and increases by a given increment. The increment is assumed to be 1, unless listed.

$$\text{seq( }\textit{expression, index, start, stop, [increment]}\text{ )}$$

As an example, we create a list of squares from 0 to 16. Figure 13.3 shows a list of squares generated with the expression listed directly in the sequence command. The last frame show how to use a predefined function, $Y_1 = x^2$ to obtain the same result and an alternate syntax to show that the index is a dummy variable.

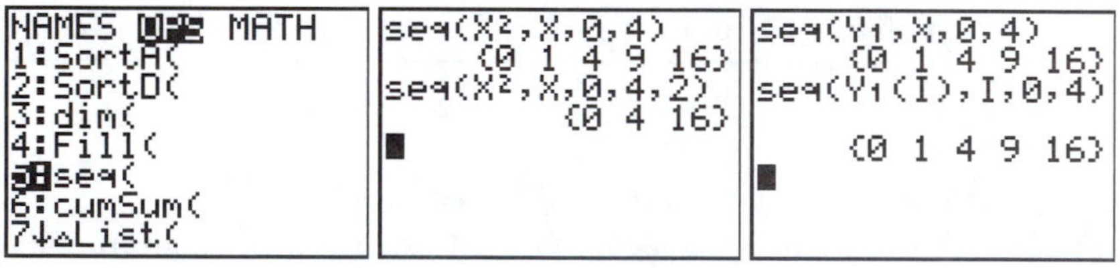

```
NAMES OPS MATH      seq(X²,X,0,4)         seq(Y₁,X,0,4)
1:SortA(               {0 1 4 9 16}          {0 1 4 9 16}
2:SortD(            seq(X²,X,0,4,2)       seq(Y₁(I),I,0,4)
3:dim(                 {0 4 16}
4:Fill(            ■                        {0 1 4 9 16}
5:seq(                                    ■
6:cumSum(
7↓ΔList(
```

*Figure 13.3 Using 2nd_LIST OPS seq() to create a list of square values and then using the function $Y_1 = X^2$ to create the same list in two ways.*

## Summing sequences to create left- and right-hand sums

The previous examples created simple lists. To sum such a list, we can use LIST MATH 5:sum(. The translation from mathematical symbols to TI commands is

$$\sum_{x=0}^{4} x^2 \Rightarrow \texttt{sum(seq(X\^{}2,X,0,4,1))}$$

A common calculus task is to form the left- and right-hand sums for a function over an interval that has been divided into *n* subintervals. This is slightly more complicated than what we have done, but it is just a sum of terms formed by a function value times the length of a interval subdivision. Geometrically, it is the sum of areas of a set of rectangles that are $f(x)$ high, by $\Delta x$ wide. That is, they approximate the area under the function's graph. In Chapter 16 we use a program that draws the rectangles and calculates left- and right-hand sums, but in this chapter we just want to understand how to build these sums.

**The left- and right-hand sums: `sum(seq(…))`**

We break this task into four parts
- Define $Y_1$, the function whose values we sum,
- Define $Y_2$ and $Y_3$ with the left- and right-hand sum formulas
- Set the interval and number of subdivisions, and calculate the subdivision length
- Evaluate the sums as $Y_2$ and $Y_3$

There is a bit of a trick here. By storing the sum formulas in the function definition of $Y_2$ and $Y_3$, we can preset all the variables used in these formulas and then the sum is the value of the function. The functions $Y_2$ and $Y_3$ are defined in Figure 13.4 with the following general format

$$\texttt{sum(seq(}\ f(x_{index})*\Delta x, \ index, \ start, \ stop)\texttt{)}$$

The left-hand sum, $Y_2$, starts at the left of the first interval so its *x*-values are $A$, $A + D$, $A + 2D$, etc. where $A$ is the left endpoint and $D$ is the length of the subinterval. For the right-hand sum, $Y_3$, we start at $A + D$ (with $I = 1$) and go until $I = N$, where $N$ is the number of divisions. First enter the function $Y_2$ and move the cursor to $Y_3$. Use 2nd_RCL $Y_2$ and the $Y_2$ definition is pasted into $Y_3$. Change the start and stop values of the index to create the right-hand sum.

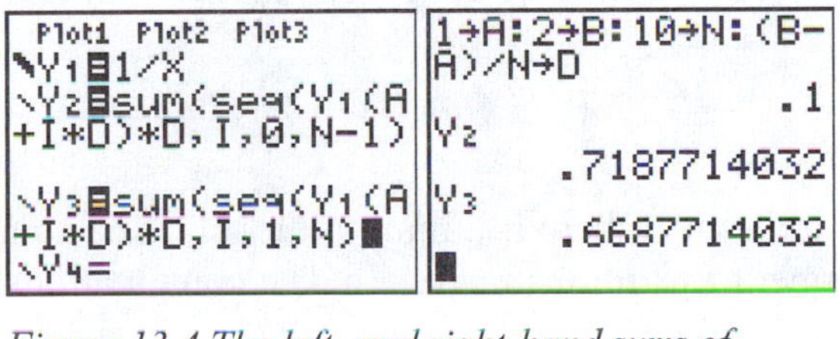

*Figure 13.4 The left- and right-hand sums of $f(x) = 1/x$ over $1 \le x \le 2$ with 10 subdivisions.*

The next step is to define the variables in the sum formula. This would run off the screen and make viewing difficult so we the colon feature which allows us to enter several commands on the same line all at once. When a colon command is used only the last command value is shown; see the .1 as the value of D. This provides a reality check on the values you entered. To repeat find a sum with a greater N, just use 2nd_ENTRY to repeat the previous commands and edit the long command line. The final step is easy just request the value of $Y_2$ and $Y_3$.

*Tip:*  Store the sum formulas out of the way in $Y_8$ and $Y_9$. Deselect them for graphing since they may appear on the screen as horizontal lines if their constant value is between Ymin and Ymax. Functions that are turned off still are active for calculations.

# Negative values in the sum

We can use our long entry line to find left-hand sums for the function $f(x) = \sin(x^2)$ on the interval $0 \le x \le \sqrt{2\pi}$ and then on the shorter interval $0 \le x \le \sqrt{\pi}$. The graph in Figure 13.5 shows that the function is negative from $\sqrt{\pi}$ to $\sqrt{2\pi}$. The fact that that the sum on the longer interval $0 \le x \le \sqrt{2\pi}$ is less than the sum on the subinterval $0 \le x \le \sqrt{\pi}$ is understandable since we are adding the negative values on the subinterval $\sqrt{\pi} \le x \le \sqrt{2\pi}$. This shows that we need to be careful when interpreting these sums as areas.

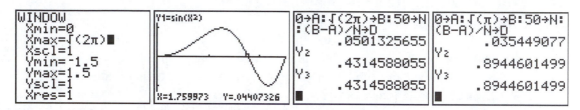

*Figure 13.5 The total sum over an interval may be less than the sum over a subinterval if the function is negative. By chance the left-hand and right-hand sums are the same.*

> *Tip:*    Use 2nd_ENTRY for deep recall of the set up line.

# Approximating area using the left- and right-hand sums

By increasing the number of partitions, the left- and right-hand sums may approach a limit which we interpret as the (signed) area under the function's graph. We write this as

$$\int_a^b f(x)\,dx = \lim_{n \to \infty} \sum_{i=1}^{n} f(x_i)\Delta x$$

As an example, let's see if there is a limit to the left hand sums of $\sin(x)$ over the interval $0 \le x \le \pi$ as the number of partitions increases. The computations with $N = 10$, 25, and 50 in Figure 13.6 suggest that the sums approach 2. We later confirm that $\int_0^\pi \sin(x)\,dx = 2$ by using the Fundamental Theorem of Calculus.

```
"sin(X)"→Y1              0→A:π→B:25→N:(B-   0→A:π→B:50→N:(B-
            Done         A)/N→D             A)/N→D
0→A:π→B:10→N:(B-                .1256637061        .0628318531
A)/N→D                   Y2                 Y2
        .3141592654              1.997367413        1.999341983
Y2
        1.983523538
```

*Figure 13.6 The left-hand sum of the sine function on the interval $0 \le x \le \pi$ approaches the value 2 as the partition increases from 10 to 25 to 50. Note the home screen method of defining Y₁.*

# CHAPTER FOURTEEN

# THE DEFINITE INTEGRAL

We saw in the previous chapter how to calculate left- and right-hand sums which approximate the signed area under a curve. The definite integral is defined as the limit of the left-hand (or right-hand) sum as the number of partitions $n$ goes to infinity. Thus, each definite integral is a specific real number, and the TI calculates this value. (Well, almost — it calculates an approximation that is generally reliable.) The definite integral is evaluated as a number, but defining its upper limit as a variable creates a new function.

## The definite integral from a graph

By displaying a graph and using the 2nd_CALC menu, we find the definite integral of a function *and* see its graphic representation.

Let's start with $y = 2\sin(x)$ and graph it in the ZTrig window. Press 2nd_CALC to see the integral option 7:∫f(x)dx. You are prompted to set the lower limit and then the upper limit. These limits are set in the same way that you have already set bounds in the 2nd_CALC options. Remember that the bounds must be within Xmax and Xmin.

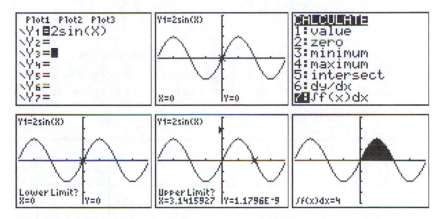

Figure 14.1 The value of a definite integral found from the graph by using
2nd_CALC. Area is a shaded on the graph. (ZTrig window.)

The numeric answer can be interpreted as a measure of shaded area under the function's graph. It may surprise you that the value is exactly 4, but notice that $Y_1$ is twice the sine function and recall from the last chapter that the left- and right-hand sums of the sine function converged to 2 over this interval.

*Tip:* If there is more than one function graphed then you must select one function before using the 7:∫f(x)dx command, otherwise it uses the first one selected.

## The definite integral as a number

From the home screen, the TI command for the definite integral is in the MATH menu, 9:fnInt(, and has the general format:

fnInt(*function, variable, lower, upper*)

For example, the translation from the mathematical symbols of the previous example to this TI command is

$$\int_0^\pi 2\sin(x)\, dx \Rightarrow \text{fnInt(2sin(X), X, 0, }\pi\text{)}$$

See the last frame in Figure 14.2.

## Facts about the definite integral

Four definite integral facts are illustrated using simple functions and windows. You are encouraged to change the function and window to make it more exciting. In each of the following examples, we show the result graphically and then, in the final frame, we show the numerical rendition of the same result on the home screen. It is important that you feel comfortable using both methods of finding the definite integral.

> *Tip:*   After a graph with shading has been displayed, it is usually desirable to clear the screen before the next graphing. This is done by pressing 2nd_DRAW, 1:ClrDraw. Using a ZOOM menu selection also gives a fresh graph. This is handy when your window is ZStandard, ZDecimal, ZTrig, or ZPrevious.

### Reversed limit integrals are the negative of one another

Unlike most settings where error messages are given whenever Xmin > Xmax or Left Bound > Right Bound, the Upper Limit and Lower Limit can be in either order. The result of an order reversal is a sign change of the value.

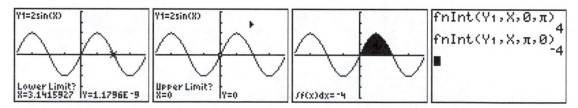

*Figure 14.2 Limit reversal changes the sign of the definite integral.*

### The intermediate stop-over privilege

The definite integral can be calculated as a whole from the lower to upper limit, or it can be calculated in contiguous pieces. This can be thought of as a plane fare where the charge is the same whether you fly non-stop or have an intermediate landing. Figure 14.3 shows that we get the same answer dividing our example function over two particular subintervals.

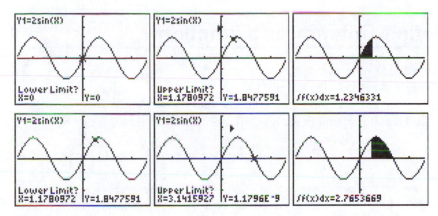

*Figure 14.3 The definite integral found in two pieces, with 7:∫f(x)dx. Notice that the sum of the two answers is 4.*

## The definite integral of a sum is the sum of the individual integrals

In Figure 14.4, we set Y₁=2sin(X), Y₂=X. and Y₃=Y₁+Y₂. We find the sum of the two definite integral values for Y₁ and Y₂ is equal to the definite integral of the sum function, Y₁+Y₂. Note that the equality of the two definite integrals in the last frame is independent of Y₁ and Y₂.

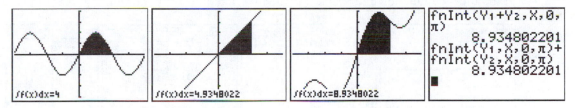

*Figure 14.4 The definite integral of a sum is the sum of the integrals.*

## Constant multiples can be factored out of a definite integral

We already saw an example of this with $\int_0^\pi 2\sin(x)\,dx = 2\int_0^\pi \sin(x)\,dx$. This example is shown in Figure 14.5 including the numeric integral in the last frame.

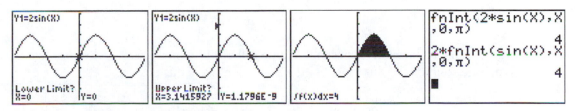

*Figure 14.5 A constant multiple of a function can be factored out of a definite integral.*

# The definite integral as a function

Recall that `nDeriv(T²,T,1)` is the derivative value at 1 and, by replacing 1 with X, we can create a derivative function, `nDeriv(T²,T,X)`. We repeat this dummy variable technique to create an integral function. We use T as the dummy variable in the examples (it could be called X and it would make no difference). For example,

$$Y_1 = \int_a^x \cos(t)\,dt = \texttt{fnInt(cos(T), T, A, X)}$$

> *Tip:*  Graphing functions defined with `fnInt()` is quite slow, even on the TI-84 models; setting `Xres` to a higher number increases graphing speed.

In Figure 14.6, we choose 0 as the lower limit, so

$$Y_1 = \texttt{fnInt(cos(T),T,0,X)},$$

and graph it using a `ZTrig` setting (and `Xres = 3`). We see that the graph of $Y_1$ looks like a sine function and we check this using a table with `TblStart = 0` and `ΔTbl = π/12`. (You could also graph the sine function and check that the two functions have the same graph.)

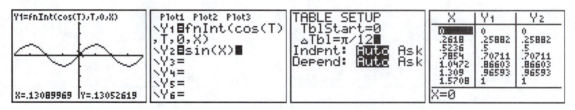

*Figure 14.6 The integral function of the cosine appears to be the sine function.*

If we change the lower limit in the `fnInt()` definition, then interesting things will happen. You can see from Figure 14.7 that the three functions are vertical shifts of one another. Thinking of the derivative as the rate of change of *y*-values, these three should have the same derivative. Each of these three functions is called an antiderivative of cos(*x*). The common notation used is

$$\int \cos(x)\,dx = \sin(x) + C$$

where *C* is an arbitrary constant. Neither the TI-84 nor TI-83 calculator can give you symbolic solutions of this type; but the TI-89 does have this power.

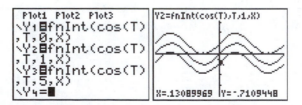

*Figure 14.7 Antiderivatives of the cosine function are of the form sin(x) + C.*

# CHAPTER FIFTEEN

# THE FUNDAMENTAL THEOREM OF CALCULUS

The Fundamental Theorem of Calculus is discussed in two forms: as the total change of the antiderivative, then as a connection between integration and differentiation.

## Why do we use the Fundamental Theorem?

The Fundamental Theorem of Calculus states

if $f$ is a continuous function and $f(t) = \dfrac{dF(t)}{dt}$, then $\displaystyle\int_a^b f(t)\,dt = F(b) - F(a)$

One reason we use this theorem is that it calculates the definite integral in a simple way. Unfortunately, this use is limited to when we know the antiderivative. For example, we previously used the Riemann sum to guess that the definite integral `fnInt(sin(x),x,Ø,π) =` 2. Using the Fundamental Theorem, we do this calculation in our head:

an antiderivative of $f(x) = \sin(x)$ is $F(x) = -\cos(x)$,

so $F(\pi) - F(0) = -\cos(\pi) - (-\cos(0)) = -(-1) - (-1) = 2$

A second reason to use the Fundamental Theorem is that it gives us an exact answer, which may be required or just plain useful. For example, when a growth factor compounds continuously, the decimal accuracy is limited to that of the calculator. This is fine when we are dealing in thousands or millions, but sometimes we have amounts that are astronomical and we want answers that are exact to whatever number of decimal places is required. Think of the value of $\pi$: it is roughly 3, or if more accuracy is needed, we might use 22/7, or better yet, 3.1459. In its exact form, the symbol $\pi$ represents full accuracy, not a decimal or fraction approximation.

## Using fnInt() to check on the Fundamental Theorem

As a simple example we calculate the definite integral value twice, once by using the Fundamental Theorem and again by using `fnInt()`, the built-in numerical integration function. Let's see this important calculation three ways, but, we now use a simpler function:

$$f(x) = \int_0^2 x^2\,dx.$$

First, graph the function Y₁=X² (window: -4.7 ≤ *x* ≤ 4.7 and -10 ≤ *y* ≤ 10, Xres=1) and use 2nd_CALC 7:∫f(*x*)dx. This is shown in Figure 15.1. Now, on a fresh home screen, use fnInt() and then use the Fundamental Theorem with $F(x) = x^3/3$ to evaluate the integral numerically as $F(2) - F(0)$. Luckily we got the same answer all three ways, even though the exact result, 8/3, is rounded. Use ▸Frac from the CATALOG to get an exact answer.

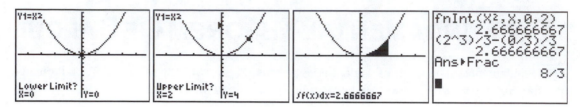

*Figure 15.1 Three different ways to calculate the definite integral.*

## The definite integral as the total change of an antiderivative

Let's look at a second example where we find an exact answer. Consider a savings account into which you put a dollar every hour. What is its worth in 20 years if it is compounded continuously at a 10% annual rate? This is a thinly disguised definite integral. First, whatever we deposit needs to be expressed in an annual amount so that all our rates are annual. Call this amount $P=365*24$ (ignoring leap years). Deposits are so frequent that we consider the rate to be continuous. The future balance in ten years is then given by the definite integral

$$\int_0^{20} P e^{0.1(20-x)} dx$$

In Figure 15.2, the fnInt() function gives a value of over half a million dollars. This is probably good enough in this case, but the calculator answer does have limited accuracy. We now use the Fundamental Theorem to write an exact answer (correct to an infinite number of digits). This is sometimes called *closed form*. We find the antiderivative function

$$F(x) = Pe^2\left(-\frac{e^{-0.1x}}{0.1}\right)$$

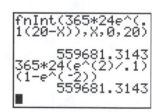

and then evaluate it at the upper and lower limits:

$$F(20) - F(0) = Pe^2\left(-\frac{e^{-2}}{0.1}\right) - Pe^2\left(-\frac{e^{-0}}{0.1}\right) = P\left(\frac{e^2}{0.1}\right)\left(1 - e^{-2}\right)$$

When written in terms of *e*, the expression has full accuracy. The two answers are compared in Figure 15.2. Knowing the closed form solution allows us to accurately find the future value with large *P*, although we might need a calculator with more internal digits of accuracy to see the difference.

*Figure 15.2 A definite integral calculated using the built-in numeric integral approximation and the closed form given by the Fundamental Theorem.*

# Viewing the Fundamental Theorem graphically

In the above examples, the lower and upper limits were constants. Now consider the upper limit as a variable. Specifically, let the upper limit be a variable on each side of the Fundamental Theorem equation, which gives a function in terms of $x$:

$$\int_a^x f(t)\,dt = F(x) - F(a)$$

We give an example by setting up the following functions:

$$f(x) = \text{Y}_1 = \emptyset.1\text{X}^2, \text{ whose antiderivative is } F(x) = \text{Y}_3 = (\emptyset.1\text{X}^3/3)$$

Since $a$ is arbitrary, let $a = -3.5$. (This means that our summing starts at zero when $x = -3.5$.) Finally, we define the two functions we want to compare and see if they are equal:

$$\text{Y}_2 = \text{fnInt}(\text{Y}_1,\text{X},-3.5,\text{X}) \text{ and } \text{Y}_4 = \text{Y}_3(\text{X}) - \text{Y}_3(-3.5)$$

Now using a ZDecimal graphing window and the graphing styles shown in the first frame of Figure 15.3, you will see that Y$_2$ and Y$_4$ have the same graph. The bubble PATH style for Y$_4$ is helpful in seeing that the two graphs are the same.

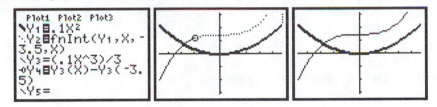

*Figure 15.3 Fundamental Theorem example with f(t) = 0.1t².*

*Tip:*  If you are verifying that two functions have the same graph, use TRACE and the up/down arrow to move back and forth between functions. Another method is to turn on PATH style (bubble icon) or BOLD style for the second function.

*Tip:*  Graphing with fnInt as part of a formula is very slow. Patience is required, especially for TI-83 users. Speed is increased by the Xres setting.

### Checking on f(*x*)= e$^x$, the function that is its own antiderivative

As another example of this kind of comparison, we let $f(x)$ be the famous exponential function, whose antiderivative is itself. We check that the two sides of the Fundamental Theorem equation are graphically equal by defining and graphing the two functions shown in Figure 15.4.

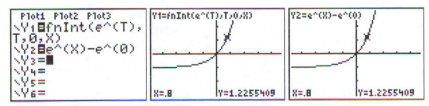

*Figure 15.4 Fundamental Theorem example with with f(t) = eᵗ. Checked by TRACE.*

# Comparing nDeriv(fnInt(...)...) and fnInt(nDeriv(...)...)

What happens when you find the derivative of the integral function? It should not be too surprising that you get the original function back. Consider $g(x) = \sin(x)/x$, whose domain is all non-zero real numbers.

In the graph of Figure 15.5, it looks like $Y_1$ and the bold style $Y_2$ match perfectly as $Y_2$ is seen tracing over $Y_1$. However, there is so much numerical action taking place in the calculation of $Y_1$ that these values are just close approximations. This can be seen in a table if you highlight a $Y_1$ entry and compare it to the corresponding $Y_2$ value: they are not exactly the same. In this case we believe the graph.

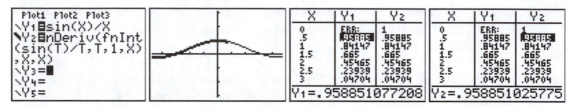

*Figure 15.5 Graph and table for* `nDeriv(fnInt(...)...)` *with closer inspection of two table entries that should be equal but differ slightly because of approximation errors.*

Let's reverse the order and integrate the derivative function. You might think this equals the original function but as you can see in Figure 15.6, you get the original function plus some constant (which depends on the lower limit you enter).

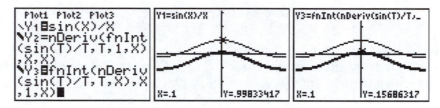

*Figure 15.6 Graph showing that* `fnInt(nDeriv(...)...)` *differs from the original function by a vertical shift of -sin(1) = -0.84147...*

# CHAPTER SIXTEEN

# RIEMANN SUMS

In Chapter 13 we introduced the right- and left-hand sums to approximate the definite integral. In this chapter we use a program to simplify explorations of these and other types of sums. We add the capability to graphically view the subdivision areas that sum to make the approximation.

## A few words about programs

This marks our first use of a stored program, so perhaps an introduction is in order. A program is a set of commands that are performed in a prescribed order. The order is normally the sequential list of commands, but there are techniques to alter that order. A program is written by pressing PRGM, arrowing to NEW, and pressing ENTER (see Figure 16.1). You then enter a name and the sequence of commands that are desired. Commands are entered from menus by pasting; most contain lower case letters, which make them easy to spot in the program listing. What is not always so easy is to know where to find the command to be paste; remember that the 2nd_CATALOG option is a complete list. Even symbols are obtained by arrowing up from abs(, (the first A listing.) Use 2nd_QUIT to exit the program editing mode.

*Figure 16.1 The* PRGM *menu and creating a new program. Alpha mode on.*

A program is activated — the more common term is *run* or *executed* — by pressing PRGM, selecting the program name, and then pressing ENTER. Step-by-step instructions can be found in the programming chapter of the *Guidebook*.

Entering a program from the printed page is quite tedious and you commonly expect to make a few errors that are only discovered when executing the program. However, once a program is correctly entered in a TI calculator, it can be transferred to other TI calculators using the built-in 2nd_LINK commands. In a classroom, it is typical that a program is verified by the instructor and distributed to the class using 2nd_LINK. You can download additional programs from the Web; this is outlined in the Appendix.

*Tip:* Programs are run from the home screen. Be sure the command line is clear before starting to execute a program.

### The program pasting error

The most common error when you first use programs is to forget that programs are pasted onto the home screen and then executed. For example, you might be in the middle of creating a command to sum a sequence when you remember that you have a program to do it. Pressing PRGM and selecting the program name RSUM pastes the command PrgmRSUM at the end of the line you were on. Some kind of error message results because the PRGM command was not on its own line.

*Figure 16.2 Pasting a command to execute a program causes an error if the home screen cursor is not on a new line.*

# Using the RSUM program to find Riemann sums

Two programs to automate the Riemann sum process are given at the end of this chapter. Before using them, you must define the function in Y₁ and set the window to have Xmin be the left endpoint and Xmax be the right endpoint of the desired interval. See Figure 16.3.

Figure 16.4 shows the program prompts you for the number of subdivisions (partitions) of the interval and then displays five different kinds of Riemann sums. The

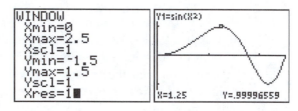

*Figure 16.3 Set Y₁ and the window before using the RSUM program.*

value labeled NINT is fnInt(Y₁,X,Xmin,Xmax) is the target, but remember that it too is only a numerical approximation. SIMP stands for the value using Simpson's method. This is a weighted average of the two previous results, specifically the sum of twice the midpoint value and the trapezoid value all divided by three; this value is consistently close to the 'true' value for small partitions.

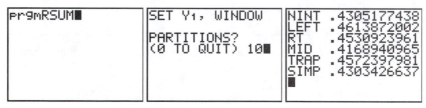

*Figure 16.4 The RSUM program gives six numeric approximations of the definite integral of Y₁ from Xmin to Xmax.*

In Figure 16.5, the graphic representations of four Riemann sums are shown. The program GRSUM has a simple menu to choose one of the four sums to graph. After a graph is displayed press CLEAR, ENTER and the menu reappears for another choice.

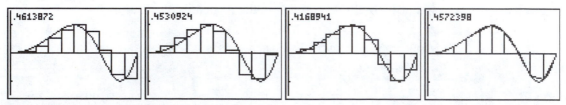

*Figure 16.5 Graph of the Riemann sum: left, right, midpoint and trapezoid, using 10 partitions.*

# The RSUM program

The programs listed below take a considerable amount of time to enter, but once entered, they can be passed to other TI calculators of the same type. (It is in the public domain.) It can also be stored on a computer using TI Connect or the older TI Graph Link.

### RSUM program listing

```
ClrHome
Disp "SET Y₁, WINDOW"
Disp ""
Disp "PARTITIONS?"
Input "(Ø TO QUIT) ",N
If N<1
Stop
ClrHome
(Xmax-Xmin)/N→D
Disp "NINT"
fnInt(Y₁,X,Xmin,Xmax)→A
Output(1,6,A)
Disp "LEFT"
sum(seq(Y₁(Xmin+I*D),I,Ø,N-1,1))*D→L
Output(2,6,L)
Disp "RT"
sum(seq(Y₁(Xmin+I*D),I,1,N,1))*D→R
Output(3,6,R)
Disp "MID"
sum(seq(Y₁(Xmin+I*D),I,.5,N,1))*D→M
Output(4,6,M)
Disp "TRAP"
(L+R)/2→T
Output(5,6,T)
Disp "SIMP"
(2M+T)/3→S
Output(6,6,S)
```

# The GRSUM program

For faster entry, you can use a copy technique on repetitive code. When you reach the section: `If A=1:Then`, enter it as a new program, G, and use `2nd_RCL PRGM EXEC G` to paste in four copies of the section, then lightly edit the other three where they differ. Then delete G.

### GRSUM program listing

```
ClrHome
Disp "SET Y₁, WINDOW"
Disp ""
Disp "1=LEFT"
Disp "2=RIGHT"
Disp "3=MID"
Disp "4=TRAP"
Input A
ClrDraw
FnOff :FnOn 1
1Ø→N
(Xmax-Xmin)/N→D
```

```
sum(seq(Y₁(Xmin+I*D),I,Ø,N-1,1))*D→L
sum(seq(Y₁(Xmin+I*D),I,1,N,1))*D→R
sum(seq(Y₁(Xmin+I*D),I,.5,N,1))*D→M
(R+L)/2→T

If A=1:Then
For(I,Ø,N-1)
Xmin+I*D→X
Y₁(X)→Y
Line(X,Ø,X,Y)
Line(X,Y,X+D,Y)
Line(X+D,Y,X+D,Ø)
End
Text(1,3,L)
End

If A=2:Then
For(I,Ø,N-1)
Xmin+I*D→X
Y₁(X+D)→Y
Line(X,Ø,X,Y)
Line(X,Y,X+D,Y)
Line(X+D,Y,X+D,Ø)
End
Text(1,3,R)
End

If A=3:Then
For(I,Ø,N-1)
Xmin+I*D→X
Y₁(X+D/2)→Y
Line(X,Ø,X,Y)
Line(X,Y,X+D,Y)
Line(X+D,Y,X+D,Ø)
End
Text(1,3,M)
End

If A=4:Then
For(I,Ø,N-1)
Xmin+I*D→X
Y₁(X)→Y
Line(X,Ø,X,Y)
Line(X,Y,X+D,Y₁(X+D))
End
Line(Xmax,Y₁(Xmax),Xmax,Ø)
Text(1,3,T)
End
```

# CHAPTER SEVENTEEN

# IMPROPER INTEGRALS

In this chapter we look at two different problems encountered when using the integral in a wider setting. These special cases involve infinity and are called improper integrals. First, we see that the limit of integration can be infinite. Second, we see that integration is sometimes possible even when the integrand function itself has infinite values.

## An infinite sum with a finite value

On a first take, you might think that any positive function that goes on forever must have an infinite definite integral. Let's try a thought experiment. Suppose you decided to go on a diet and every day you cut your chocolate chip cookie consumption in half. How many cookies would you need for your lifetime? (or for eternity?) Visualize the cookie: the first day you eat half, the next day a half of a half (a quarter), and so on. Because each day you only eat half of the remaining cookie, you never finish it: one cookie lasts a lifetime! This is represented by using the tricky function $Y_1 = (1/2)^{int(x)}$. The int() function is found in MATH NUM 5:int( and gives the integer part of a number so that the function is $\frac{1}{2}$ until 1, then $\frac{1}{4}$ until 2, etc. The table in Figure 17.1, gives the values of the integration, and suggests that the total area under the graph is 1. The table setup is set to Ask so that the X values can be entered and the Y values are only calculated when the cursor is in a cell and ENTER is pressed. Finding values when fnInt is in the function definition is time consuming.

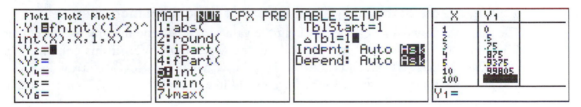

*Figure 17.1 The sum of cookie halving is one.*

## An infinite limit of integration

Now that we know a definite integral can have an infinite limit and a finite sum, we investigate which functions have this property. Let's compare the three functions

$$Y_1 = \frac{1}{x}, \; Y_2 = \frac{1}{x^3} \text{ and } Y_3 = \frac{1}{x^{1/3}}$$

as $x$ goes from 1 to infinity and see if we get a graphical and numerical hint about which might have a finite sum. To have a finite sum, a continuous function must approach zero. We see in Figure 17.2 that all three functions do approach zero as $x$ gets large, and $Y_2$ does so the fastest.

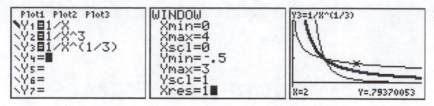

*Figure 17.2 Comparing three functions as x goes to infinity.*

## Graphing the integral with the upper limit as variable

> **Tip:**   When using the `fnInt()` function, expect long calculation times. If the wait is too long to bear, press `ON` to break.

In a different approach we graph the values of the definite integral as $x$ becomes large. Unfortunately, this takes several minutes. After you set it up to graph, take a long break while it works. If Figure 17.3 we see that $Y_1$ and $Y_3$ do not appear to have a finite sum, but $Y_2$ becomes a constant of 0.5. Who knows, maybe $Y_1$ and $Y_3$ become constant for very large $x$? Actually, we know that the first integral is defined to be $\ln(x)$, so it does not converge to any finite number. In Figure 17.3, the bold graph of $Y_1$ looks like the graph of $\ln(x)$.

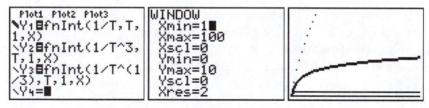

*Figure 17.3 Graphs of the integral functions as the upper limit gets large $Y_2$ converges; the others diverge. (Note Xmin=1.)*

## Making a table of the integral values as the upper limit increases

The table mode works well for checking the values of a definite integral as $x$ becomes large. We again set both `Indpnt:` and `Depend:` to the `Ask` setting. We fill this column with larger and larger values (we have freedom to choose each $x$-value.)  Next we arrow to a $Y_i$ column and press `ENTER` in any location where we would like to evaluate $Y_i(X)$. You can use the right arrow to reach $Y_3$ and find its values.

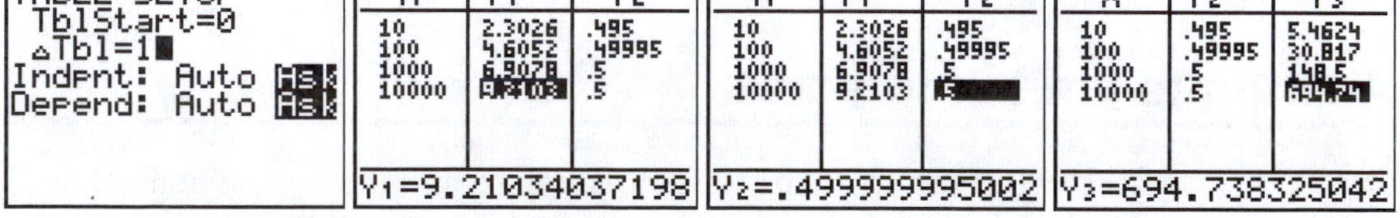

| TABLE SETUP | X | Y₁ | Y₂ |   | X | Y₁ | Y₂ |   | X | Y₂ | Y₃ |
|---|---|---|---|---|---|---|---|---|---|---|---|
| TblStart=0 | 10 | 2.3026 | .495 |   | 10 | 2.3026 | .495 |   | 10 | .495 | 5.4624 |
| △Tbl=1■ | 100 | 4.6052 | .49995 |   | 100 | 4.6052 | .49995 |   | 100 | .49995 | 30.817 |
| Indpnt: Auto **Ask** | 1000 | 6.9078 | .5 |   | 1000 | 6.9078 | .5 |   | 1000 | .5 | 148.5 |
| Depend: Auto **Ask** | 10000 | **9.2103** | .5 |   | 10000 | **9.2103** | **.5** |   | 10000 | .5 | **694.74** |
| | Y₁=9.21034037198 | | | Y₂=.499999995002 | | | | Y₃=694.738325042 | | | |

*Figure 17.4 Table values of the integral functions as the upper limit gets larger.*

> **Tip:**   The truth is that $\int_1^\infty \frac{1}{x^p}\, dx$ converges only when $p > 1$.

# The convergence of $\int_0^\infty \frac{1}{e^{ax}}\,dx,\,(a > 0)$

It is simple to show analytically that $\int_0^\infty \frac{1}{e^{ax}}\,dx = \frac{1}{a}$ for $a > 0$. Let's try this out using the table values with $a = 5$. In Figure 17.5 the integral is exactly approximated to be 0.2 with an upper limit of 10, so it should continue to calculate that value for higher values, right? Indeed it should, but for larger values of the upper limit, the approximation becomes zero! A moment's thought tells you why it is wrong: the dreaded sampling problem lurks behind this error. As the interval becomes larger, the calculator no longer samples near the 0 and thus misses the values of the function that contribute to give 0.2.

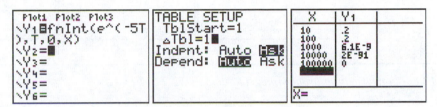

Figure 17.5 Table values of the integral functions as the upper limit gets larger and the accuracy deteriorates.

*Tip:* In 2nd_TBLSET, if you have Indpnt: set on Ask and Depend: set on Auto, then it calculates all the y-values without requesting them individually.

# The integrand goes infinite

The second way an integral is improper is when the integrand itself has at some point an infinite value (a version of being undefined). This condition is hidden because there is no ∞ symbol to alert you. It is a good habit to graph a function before finding the definite integral. The graph alerts you to potential problems like the integrand being undefined (and tending to positive or negative infinity). In Figure 17.6 the calculator warns you that it cannot approximate the integral; in such cases, you should suspect that the integral does not converge and look at its graph.

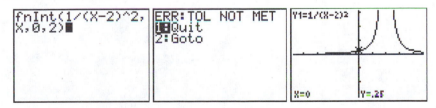

Figure 17.6 An error message for a divergent integral using `fnInt()`.

## Results we can trust

From the graph screen, if you use the CALC 7:∫f(x)dx option to evaluate an improper integral, then calculation time is extended and in some cases you receive an error message.

This could signal that the integral diverges, but you should be careful since this is not a reliable test. The examples in Figures 17.7 and 17.8 do give accurate results.

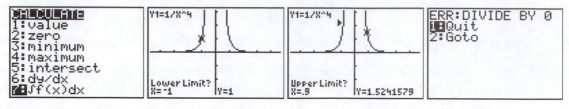

*Figure 17.7 An error message alerts to the divergent integral using 7:∫f(x)dx.*

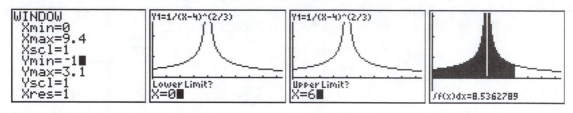

*Figure 17.8 A convergent improper integral found using 7:∫f(x)dx.*

## Results we can't trust

The calculator is a small technological device and is not completely reliable in handling improper integrals. Calculation time is long for improper definite integrals and the result are mildly undependable. The example in Figure 17.9 shows an error message, suggesting that the integral diverges. However when we know from the analytic results that since the power, 5/6, is less than one, then it converges. A calculus teacher can easily make up examples where the calculator misleads you. Would someone really do that? Yes.

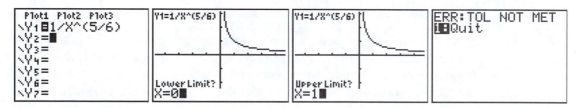

*Figure 17.9 An error messages is shown using 7:∫f(x)dx even though the improper integral is convergent. (ZDecimal window)*

# CHAPTER EIGHTEEN

# APPLICATIONS OF THE INTEGRAL

We look at applications of the integral. These typical examples give only a flavor of the extensive applications of the integral.

## Geometry: arc length

There are some calculators that calculate arc length as a command. However, the TI-84/83 requires entry of the following calculus formula.

$$\text{Arc length} = \text{L} = \int_a^b \sqrt{1 + (f'(x))^2}\, dx$$

Let's calculate the arc length of the curve $y = x^3$ from $x = 0$ to $x = 5$ and compare it to direct distance from the origin to the point (5,125). This is shown in Figure 18.1. The graph's scale of the axes makes the difference between the two lengths look greater than the actual numerical difference we found.

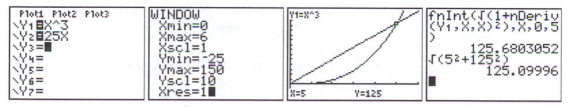

Figure 18.1 Finding the arc length of the curve $y = x^3$ from $x = 0$ to $x = 5$ and comparing it to the direct distance between (0,0) and (5,125).

## Physics: force and pressure

Pressure increases with depth so that there is more pressure at the bottom of a container than at the top. One cubic foot of water weighs 62.4 pounds and it has a force of 62.4 pounds on the base. If it is only half full then it has a force of 62.4/2 = 31.2 pounds on the base. The pressure on the base is directly proportional to the depth of the water. We also know that force is the product of pressure and area.

The difficult part of pressure, force, or volume problems is not the actual integration that is required, but setting up the integral to accurately reflect the geometry of the situation. A time-honored system is to write the pressure as a sum of forces acting on strips or slices.

### The pressure on a trough

Consider the trough shown in Figure 18.2. There are four sides that the force of water acts on. Let's tackle the easiest side first: the 3' by 14' horizontal back. We subdivide the height into pieces of length $\Delta h$, so this back side is made up of horizontal strips, each having an area of $14\Delta h$ square feet. The entirety of each strip is at the same depth, so that all along a given strip there is equal pressure, namely

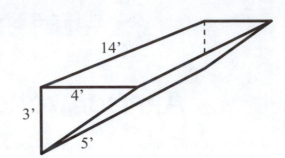

Figure 18.2 A trough with the shape of a triangular prism.

force on a horizontal strip = $(62.4h)(14\Delta h)$

(We could multiply the constants, but the calculator can do that and we choose to leave our setup in this more readable form.) The total force is the sum of all the horizontal strips. The exact total is given by the definite integral

$$\int_0^3 (62.4h)(14)\,dh$$

Next we consider the inclined side. For a $\Delta h$ (vertical height change) the actual width $w$ on the inclined side is greater than $\Delta h$. Using similar triangles, these values are related by $\Delta h/w = 3/5$. The area of the strip is $14(5/3)\Delta h$, so

force on inclined strip = $(62.4h)(14)(5/3)\Delta h$

The total force is given by the integral

$$\int_0^3 (62.4h)(14)(\tfrac{5}{3})\,dh$$

Finally we compute the force on the triangular ends. Again using similar triangles, we find the area of a strip is $(4/3)(3-h)\Delta h$ and

force on end strip = $(62.4h)(4/3)(3-h)\Delta h$

so that the third integral is

$$\int_0^3 (62.4h)(14)(\tfrac{5}{3})(3-h)\,dh$$

Now the easy part is entering these definite integrals for evaluation; see Figure 18.3.

Figure 18.3 The three definite integrals give the force on a trough.

> *Tip:* It is much safer to leave calculator results in an expanded form so that the derivation remains evident.

# Economics: present and future value

Suppose you win a 2 million dollar lottery. Before you spend the money, you are told that the money is distributed to you over the next 20 years. That is a mere $100,000 per year (before taxes). If you wanted to get an immediate lump sum, you could sell your rights to all the future payments. What is this really worth? In economics this value is called the present value, $V$. It is calculated using a fixed investment rate and an integral. If $P$ is the annual payment and $r$ is the annual investment rate for $T$ years, then the present value is

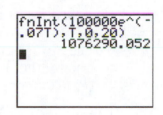

*Figure 18.4 The present value of a 2 million lottery.*

$$V = \int_0^T Pe^{-rt}\,dt$$

Suppose that the agreed investment rate was 7% for the twenty years. Then we enter the integral as shown in Figure 18.4. The present value is $1076290.05 (about half!).

### Discrete vs. continuous

The above analysis makes the assumption that the income is coming in continuously, which is not the case. If we want to calculate to the penny, we use a discrete sum for the twenty years. The point is that there is a close relationship between the integral and the sum of a sequence. The integral must be used when the income is continuous, but can be used as an approximation for a discrete sequence sum. In Figure 18.5 we see the sum, about $1,100,000, is more than the integral calculation because you receive all the $100,000 payments at the start of the year.

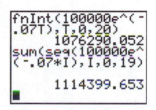

*Figure 18.5 The discrete sum is more than the integral because it is paid at the beginning of the year.*

### The future value

Since the probability of winning the lottery is essentially zero, you might want to create a jackpot for yourself by investing $100,000 a year (that's just $0.19 a minute). Your total for $T$ years is called the future value and is given by the definite integral

$$V = \int_0^T Pe^{r(T-t)}\,dt$$

Using `fnInt()`, you can find that in 20 years you would have a real jackpot worth over four million dollars.

### A note on TI financial functions

The TI-83 Plus and TI-84 have these financial functions and many others preloaded as APPS. The older TI-83 actually has all of these financial functions built in under the 2nd_FINANCE menu.

# Modeling: normal distributions

In statistics, a normal distribution has a graph that is a bell-shaped curve. Its general equation is

$$p(x) = \frac{1}{S\sqrt{2\pi}} e^{-(x-M)^2/(2S^2)}$$

where $M$ is the mean and $S$ is the standard deviation. (In a statistics course, you would use the variables $\mu$ and $\sigma$ for these values.) If $M = 0$ and $S = 1$, then the curve is called the *standard* normal curve, as shown in Figure 18.6. Carefully enter the equation $Y_1=1/(S\sqrt{(2\pi)})e^{\wedge}(-(X-M)^2/(2S^2))$, then set $\emptyset{\to}M$ and $1{\to}S$.

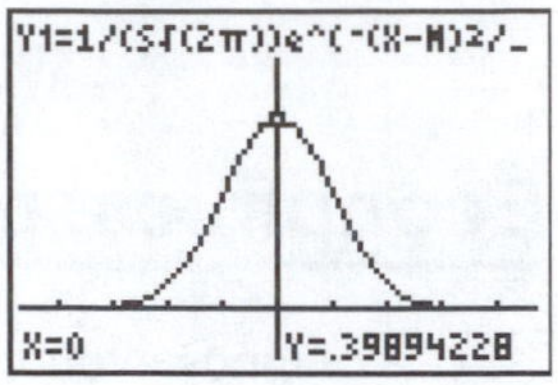

*Figure 18.6 The standard normal curve.* ZDecimal *with -0.1 ≤ y ≤ 0.6.*

## The Anchorage annual rainfall

One application of the normal distribution is to model situations where measurements are taken under conditions of randomness. For example, suppose you look at the records for annual rainfall in Anchorage, Alaska over the past 100 years. Let's simplify and say that you found the average of these averages to be 15 inches. There is another statistical indicator, called the standard deviation, telling how spread out the values are. Let's say that the standard deviation was 1. (For any list you can use the LIST MATH menu to find the mean, 3:mean, and the standard deviation, 7:stdDev.)

To estimate the percentage of the years that rainfall is between

(a) 14 and 16 inches, (b) 13 and 17 inches, and (c) 12 and 18 inches

take three integrals of the normal distribution with $M = 15$ and $S = 1$. Use CALC 7:∫f(x)dx to see from the graph in Figure 18.7 that the model predicts: (a) 68% of the years have rainfall between 14 and 16 inches, (b) 95% of the years have rainfall between 13 and 17 inches, and (c) 99% of the years have rainfall between 12 and 18 inches.

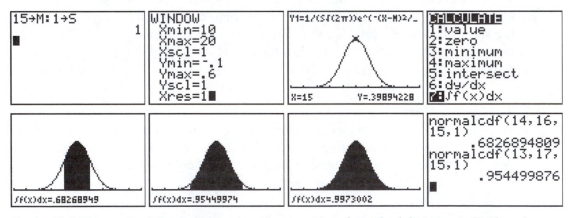

*Figure 18.7 The graph of the normal curve with mean 15 and standard deviation 1. Finding the integrals with limits (a) 14 ≤ x ≤ 16, (b) 13 ≤ x ≤ 17, and (c) 12 ≤ x ≤ 18. Checking values against built-in statistics function values for (a) and (b).*

*Tip:*    The TI has powerful statistical features that are not addressed here, such as normalcdf, which calculates the above results in the last frame of Figure 18.7. See the statistical examples in the *TI Guidebook* and/or on the TI website.

## CHAPTER NINETEEN

# TAYLOR SERIES

The use of series for approximation is shown for simple cases on the TI, but a more powerful calculator or computer is needed for any serious work. On a more powerful calculator, such as the TI-92, a Taylor series command is built in. We limit ourselves here to confirming a few well-known results. It is assumed that you know the formula for the Taylor polynomial.

## The Taylor polynomial program

In our limited scope, we derive Taylor polynomials of degree up to six. This is mainly because the TI can not nest derivative definitions. Since we need to evaluate derivatives of higher orders, we enter them into Y definitions by hand, (which means that we find them algebraically!). We use the program TAYLOR to calculate and store in $L_6$ the $f^{(n)}$(Center) values of the coefficients. The polynomial function is stored in $Y_0$ and graphed with the original function for comparison.

### TAYLOR program listing

```
ClrHome
Disp "SET:"                         Y₁(C)→L₆(1)
Disp "Y₁, WINDOW"                   Y₂(C)→L₆(2)
Disp "DERIVS IN Y'S"                If R≥2
Disp "PUTS:"                        Y₃(C)→L₆(3)
Disp "Y' VALS DERIV"                If R≥3
Disp "VALUES IN L₆,"                Y₄(C)→L₆(4)
Disp "POLY IN Y₀"                   If R≥4
Input "GO?(1=YES) ",A               Y₅(C)→L₆(5)
ClrHome                             If R≥5
If A≠1                              Y₆(C)→L₆(6)
Stop                                "sum(seq((L₆(I+1)/I!)(X-C)^I,I,Ø,R))"→Y₀
Input "DEGREE?(1-6) ",R             Func:FnOff :FnOn Ø,1
R+1→dim(L₆)                         GraphStyle(Ø,2)
Input "CENTER? ",C                  DispGraph
```

---

*Tip:* If you have function definitions in the Y's or an important list in $L_6$ that you want to save, the TAYLOR program gives you a chance to exit (Ø=N) and save them.

---

*Tip:* Change GraphStyle(Ø,2) to GraphStyle(Ø,5) if you prefer a bubble path for $Y_0$.

## Setting up before using TAYLOR

Enter the function in $Y_1$, then $Y_1'$ in $Y_2$, $Y_1''$ in $Y_3$, etc. Deselect all the functions except $Y_1$ and find a nice window for $Y_1$. Leave some room above and below the function graph so the approximation graph is seen. Check the list $L_6$ to be sure it doesn't contain data you want — it is erased when we run the program.

We use the TAYLOR program for graphing the function $y = \ln(x)$ and the Taylor polynomial of degree five at center $x = 1$. See a typical setup in Figure 19.1.

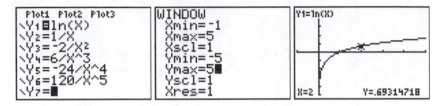

*Figure 19.1 Setting up to use the TAYLOR program.*

Run the program from a fresh line on the home screen by pressing PRGM, selecting TAYLOR from the menu, and then pressing ENTER twice. The opening screen, as shown in Figure 19.2, is just a reminder that the $Y_1$ should be the original function and $Y_2$ should be the first derivative, $Y_3$ should be the second derivative, etc. Preset the window so that it shows the function and has some extra screen space for graphing the Taylor polynomial. You will be unsure exactly where the Taylor polynomial will wander and you may need to adjust your screen later. If you have done a setup, then press 1 ENTER to continue.

Next you will be prompted to enter the degree and center. The calculator will then compute and store the derivative values, $f^{(n)}$(Center), in the list $L_6$. But note that these are not the actual polynomial coefficients until they are divided by $n!$, which is done in the function definition $Y_0$. The final prompt offers to graph your original function and the Taylor polynomial.

In Figure 19.2, we see the graphs of the function $y = \ln(x)$ and the fifth degree Taylor polynomial centered at 1. The Taylor polynomial has been graphed in bold. The approximation looks very good when $x$ is close to 1, but otherwise it is not very good, especially above $x = 2$.

*Figure 19.2 Typical screens using the TAYLOR program.*

## After TAYLOR has finished

Press CLEAR to return to the home screen and then ENTER to restart TAYLOR if you want to see another degree polynomial or to change the center.

# The Taylor polynomials for $y = e^x$

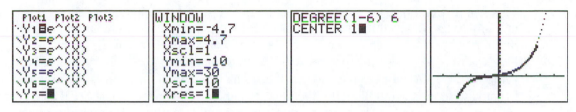

*Figure 19.3 A Taylor polynomial of degree 6 centered at 1 approximates $y = e^{\wedge}x$ when close to 1.*

Let's look at another example where knowing the higher order derivatives is easy. The function $y = e^x$ is its own derivative, so we repeat it in each $Y_i$. We find the sixth degree Taylor polynomial with center 1. The bold style graph in Figure 19.3 shows the approximation is quite bad before $x = -1$ but a good fit to about $x = 4$. We can only expect a polynomial to do a good job locally since $y = e^x$ approaches 0 as $x \to -\infty$ and no (nontrivial) polynomial does that.

# The interval of convergence for the Taylor series of the sine

In the following example, we find the Taylor polynomials for $y = \sin(x)$ centered at 0. Enter the five consecutive derivative functions and set the window with ZOOM ZTrig. In Figure 19.4, only the graphs for degrees 1, 3, 5 are shown because the graphs for degrees 2, 4, 6 (respectively) are the same. This is because the even powers have zero coefficients, as seen if you look in $L_6$. Looking at the three graphs, we see that as the degree gets larger, the approximations are close on a wider and wider interval. The interval of convergence for the third degree Taylor polynomial appears to be $-\pi/2 < x < \pi/2$, and the fifth degree is even wider.

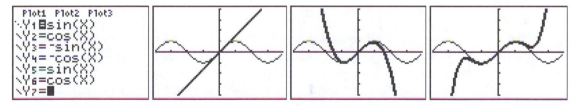

*Figure 19.4 The graphs of the Taylor polynomials of degree 1,3,5 are shown in bold. (ZTrig window)*

The fact that we are approximating trigonometric and exponential/logarithmic functions is an example of the fact that any function that is infinitely differentiable can be locally approximated by polynomials. This is important because computers and calculators are masters at polynomial evaluation; after all, a polynomial is just addition and multiplication.

Evaluating a Taylor polynomial at $x = 1$ is the same as summing the coefficients. As the degree of the Taylor polynomial goes to infinity, the value at $x = 1$ remains a finite value because of the $n!$ term in the denominator of the Taylor coefficients. Another way of saying this is that the series of Taylor coefficients converges. This leads to a more general question of how we know if a series converges.

*Tip:*   When you finish using the TAYLOR program the $Y_0$ function is selected. Turn it off before further graphing since it produces an additional unrelated graph.

# How can we know if a series converges?

This is a difficult question, but we have the ratio test and the alternating series test to help us in many cases. If asked for proof, you need to use an analytical argument, but a graph or table often tells you whether to pursue a proof of convergence or divergence. Since you are investigating a series, the $n$th term is given symbolically (or else you need to write it symbolically). In the examples below, the sigma summation notation is given, but if a series is listed without a symbolic term, then the first order of business is to write it as a sigma sum. This symbolic term is entered in the seq() command. Use Y₁ = sum(seq()) to either graph or make a table of values from which to form an opinion about whether the series converges or diverges.

**The harmonic series:** $1 + \dfrac{1}{2} + \dfrac{1}{3} + \dfrac{1}{4} + \cdots = \displaystyle\sum_{n=1}^{\infty} \dfrac{1}{n}$

We translate the partial sums into a TI function where the $x$th term is the final one:

$$Y_1 = \text{sum(seq(1/N,N,1,X,1))}$$

A good window for this kind of series investigation is Xmin=0 and Xmax=94. This means that the $x$-values are sampled at integer values. By the 94th term, the long term behavior is usually evident. See Figure 19.5. The graphing gets progressively slower as $x$ gets larger, but you can always press ON to break at any time.

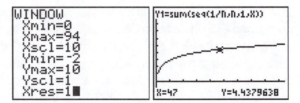

*Figure 19.5 Suspecting divergence from a graph.*

Table values are best found with Indpnt: and Depend: set to Ask. Otherwise, it takes an inordinately long time to display one screen of table values. (You often get nothing but an error message because the calculator's capacity has been exceeded.) Figure 19.6 shows some table computations. You might ask why we don't just find the 1000th partial sum. This is a good idea in theory, but in practice it is too much for this calculator: an INVALID DIM error is shown.

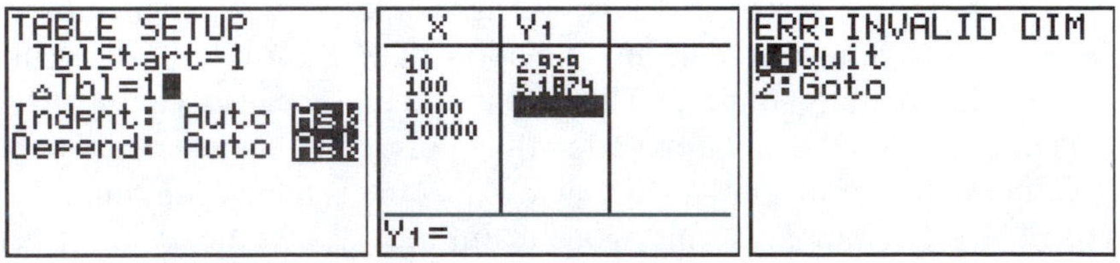

*Figure 19.6 Suspecting divergence from a table, but getting an error message by going beyond the calculator's capacity.*

**The alternating harmonic series:** $1 - \dfrac{1}{2} + \dfrac{1}{3} - \dfrac{1}{4} + \cdots = \displaystyle\sum_{n=1}^{\infty} (-1)^{n-1}\left(\dfrac{1}{n}\right)$

Often series alternate positive and negative terms. This sign switch is cleverly written in sigma notation using a power of negative one. Since every other term is subtracted, we expect the range of the partial sums to not be as large as it is for the harmonic series. Indeed, we find that this series converges.

> **Tip:** When writing a sigma sum symbolically, it is easy to make small errors. Before graphing or making a table, you might want to test a small list of terms to verify that your notation is correct.

We define and graph the partial sums function

$$Y_1 = \text{sum (seq((-1)^(N+1)/N,N,1,X))}$$

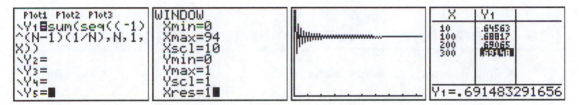

Figure 19.7 Suspecting convergence from the graph, we also see evidence for convergence in the table values.

Use $0 \le x \le 94$ so the calculations will be at integer values. Watch the graph and press ON to break when you are satisfied. We see from the graph and table in Figure 19.7 that it is a good bet that this series converges but the table shows that accuracy is slow to be obtained.

**A fast converging series:** $2 - \dfrac{2}{3} + \dfrac{2}{9} - \dfrac{2}{27} + \cdots = \displaystyle\sum_{n=0}^{\infty} (-1)^{n}\left(\dfrac{2}{3^n}\right)$

This series is related to the exponential function $f(x) = \dfrac{2}{3^x}$ and we know its values get close to zero as $x$ gets large. Figure 19.8, shows

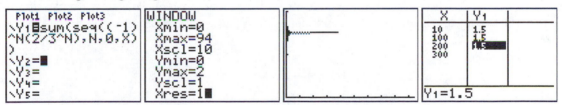

Figure 19.8 Suspecting convergence, we BREAK the graphing before complete. The table shows a fast converging series.

> **Tip:** Test your symbolic expressions on the home screen for proper use of parentheses, negative/subtraction signs, and syntax. Then use 2nd_ENTER to paste them in a Y= definition.

# CHAPTER TWENTY

# GEOMETRIC SERIES

In the last chapter, we found Taylor polynomials by starting with a function and forming the successive sums to find a series representation. In some cases, there is the reverse situation: the series is known, but the function is not. We now consider finding the function from a given series. The graphing calculator is of little help in writing a general formula for a sum, but it helps clarify the generalized sum and verify it with finite sums.

## The general formula for a finite geometric series

A finite geometric series has the form

$$a + ax + ax^2 + \cdots + ax^{n-1} + ax^n$$

Note that the coefficient $a$ is the same for each term. The closed form sum is

$$S_n = \sum_{i=1}^{n} ax^{n-1} = a\frac{(1-x^n)}{(1-x)}, \quad (x \neq 1)$$

You might think that such a sum would be so rare as to make it not worth considering, but this kind of finite series comes up in many situations.

### Repeated drug use levels

Consider a 250 mg dose of an antibiotic taken every six hours for many days. The body retains only 4% of the drug after six hours. The interesting part is that this does not say that 4% of 250 mg is left. This is only true at the end of the first six hours. At the end of the second six hours, the body retains 4% of the second dose *and* 4% of the remaining first dose. Let's make a sequence of the amount of drug in the body right after taking the $n$th dose.

$$Q_1 = 250,$$

$$Q_2 = Q_1(0.04) + 250,$$

$$Q_3 = Q_2(0.04) + 250, \text{ etc.}$$

But if we substitute lower sums and multiply, we find

$$Q_2 = 250(0.04) + 250 \text{ and}$$

$$Q_3 = 250(0.04)^2 + 250(0.04) + 250$$

In general,

$$Q_n = 250(0.04)^{n-1} + \ldots + 250(0.04)^2 + 250(0.04) + 250.$$

The $Q$'s are a finite geometric series with $a = 250$ and $x = 0.04$. In Figure 20.1, let's calculate the amount of antibiotic in the body at the time of taking the fourth pill. We use the 2nd_LIST menu to enter the sequence and see in the first frame the amount of each pill that is left. The total is found by summing the sequence. Check the answer by using the finite formula. To see consecutive finite sums of a sequence, you can define a function using `sum(seq(...))` and set up a table to show the desired values.

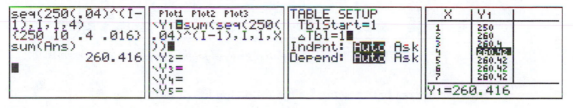

Figure 20.1 Showing the finite sum for four terms and making a table from a function.

You might observe that by that at the end of the first day (the fifth pill), your medication is close to a stable amount. A different drug might have a much higher retention level. By changing 0.04 to 0.50, in Figure 20.2, we make a table that shows a drug taking about a week to get close to its stable limit of 500 mg.

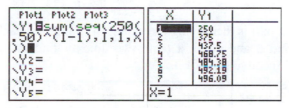

Figure 20.2 Retention amounts for a rate of 50%.

## Regular deposits to a savings account

Another direct application of the finite geometric series is the value of an investment that earns interest. Suppose that you plan for retirement by putting $1000 a year into a savings account that earns 5% annual interest. You want to know how much this is worth after the $n$th deposit. In Figure 20.3, we find values using a table and check the finite sum formula for when $n = 41$, a year you might start thinking about retirement.

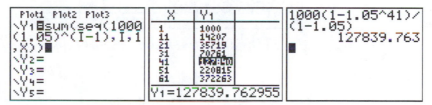

Figure 20.3 The value of a 5% savings account after 41 annual deposits.

# Identifying the parameters of a geometric series

When we use different sums of the type $a + ax + ax^2 + \cdots + ax^{n-1} + ax^n$, it is helpful to have a generalized definition of Y₁. In Figure 20.4 we do this with $a$ changed to the capital $A$ and $x$ renamed $B$. Beware that in this definition $x$ is the $n$ — there is no way to avoid this confusing assignment of variables since a Y= definition must have $x$ as the independent variable. We check our new definition with the first antibiotic retention model.

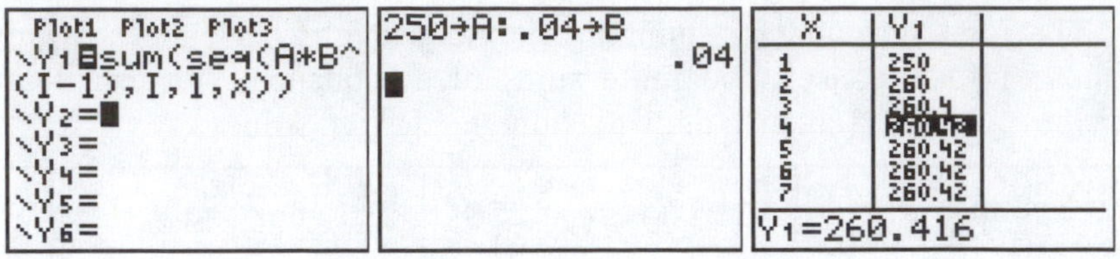

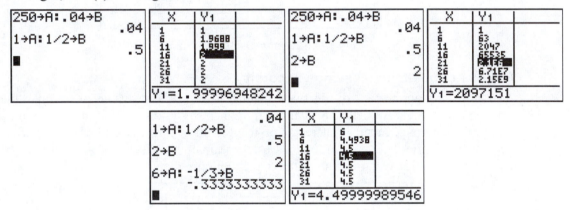

Figure 20.4 Checking to see that a general method gives the same results as before.

Now we are ready to practice on the following list of infinite series.

$$\text{(a) } 1+\frac{1}{2}+\frac{1}{4}+\frac{1}{8}+\cdots$$

$$\text{(b) } 1+2+4+8+\cdots$$

$$\text{(c) } 6-2+\frac{2}{3}-\frac{2}{9}+\frac{2}{27}-\cdots$$

The hardest part is identifying the values of $A$ and $B$ in each case. In Figure 20.5 we enter values for A and B and check a table; we quickly find that series (a) converges to 2, (b) diverges, and (c) converges to 4.5.

Figure 20.5 Use a generic $Y_1$ and set A and B to check three series for convergence.

# Summing an infinite series by the formula

We could (hypothetically) take a dose every six hours forever and the series would be infinite. We have seen infinite sums that add up to finite values, such as the 'half a cookie diet' example of Chapter 17. The infinite sum of a geometric series is given by the following formula:

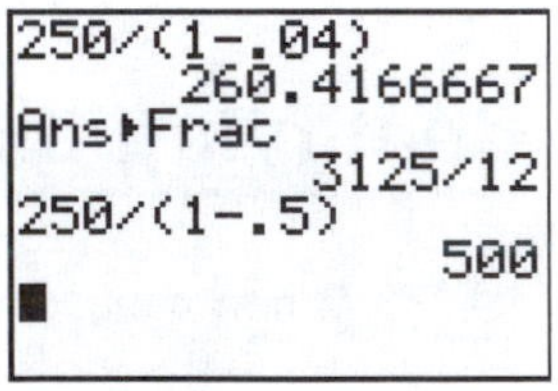

$$a + ax + ax^2 + \cdots + ax^{n-1} + ax^n + \cdots = \frac{a}{1-x}, \quad \text{for } |x| < 1$$

Figure 20.6 Infinite sums.

In Figure 20.6 we check the formula for the two drug dosage cases we investigated as finite sums. They both qualify since each $x$-value ($x = 0.04$ and $x = 0.5$) satisfies the condition $|x| < 1$. In the previous example, series (b) had $x = 2$ and did not converge.

# Piggy-bank vs. trust

Suppose that your parents are deciding on a plan to provide for your future and they have two choices:

I. Each year they put your age in dollars into a piggy bank.

II. Each year they put $3 into a savings account that earns 6% annual interest.

We already know how Plan II works: use $A = 3$ and $B = 1.06$ in the formula above. In plan I we need to look at the sum of the series $1 + 2 + 3 + 4 + \ldots + n$. This is *not* a geometric series. A clever way to consider this sum is to write it twice, once forward and once backward, as follows.

$$
\begin{array}{ccccccccc}
1 + & 2 + & 3 + & \ldots + & n\text{-}1 + & n & = & S_n \\
n + & n\text{-}1 + & n\text{-}2 + & \ldots + & 2 + & + 1 & = & S_n \\
\hline
n+1 + & n+1 + & n+1 + & \ldots + & n+1 + & n+1 & = & 2S_n
\end{array}
$$

There are $n$ of these $(n+1)$ sums, so we have $S_n = n(n+1)/2$. We use a simplified version of this formula in Y$_2$ and look at some early values in a table in Figure 20.8. It seems that the piggy bank is the better plan. But arrowing down the table we see that as you get to 66 years old, the two plans seem about even, and as you get into your seventies, plan II is much better.

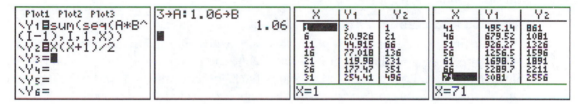

Figure 20.7 A table for the trust and the piggy bank plans.

To see the actual break-even point a graph serves us better. In Figure 20.8, a graph is used to find the intersection point. Be warned that the first graph takes time. Then finding the intersection point takes even longer. In this case it would have been faster to TRACE, since our goal is to find a year and this window will trace on integer values.

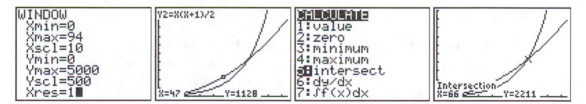

Figure 20.8 A graph for the trust and the piggy bank plans.

# CHAPTER TWENTY ONE

# FOURIER SERIES

The Taylor polynomials were good approximations to a function, but beyond the radius of convergence they were awful. As suggested by the example of the sine function in Chapter 19, Taylor polynomials are practically useless for periodic functions. We now find a set of approximating functions that globally approximate a periodic function. These global approximation functions are called Fourier approximations. First we discuss how to graph some periodic functions that are not trigonometric, including piecewise defined functions.

## Periodic function graphs

The only kinds of periodic functions we have graphed so far are trigonometric. We now introduce and graph a few other types.

### The rise and crash function

Our first example is $y = x - \text{int}(x)$, this is a linearly increasing function that falls back to 0 when it reaches a height of 1. It is defined using the function $f(x) = \text{int}(x)$, called the greatest integer function. It is in the CATALOG and in the MATH, NUM menu. Use the DOT style to graph this type of function, so you can avoid the jagged connected graph problem.

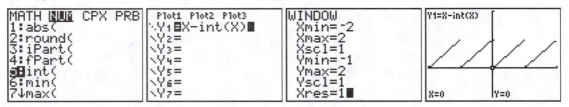

*Figure 21.1 A non-trigonometric periodic function.*

### The square wave function: using a logical expression

The square wave function is commonly used in electrical engineering to model switching, since it is either on or off. We use 1 to mean on, 0 to mean off. In this case we graph only one period of the function using the logic feature of the TI.

If we evaluate the logical expression (Ø<X), then it will be TRUE or FALSE depending upon the current value of $x$. The arithmetic value of TRUE is 1, of FALSE is 0. Thus an expression like (Ø<X) has a value of one if $x < 0$ and zero otherwise. Inequality symbols, such as < and ≤, are pasted in from the 2nd_TEST menu. We can actually make a logical expression into a function and graph it; see Figure 21.2.

If we multiply two logical expression as a product, say (Ø<X)(X≤1), then the value is 1 for $0 < x \leq 1$, but zero otherwise. If we think about the square wave function of period 2, then this product reproduces that function for one complete period $0 \leq x \leq 2$. We approximate and graph Fourier approximations of this function later in this chapter.

*Figure 21.2 A logical expression in a function definition.*

## Piecewise defined functions

The technique of putting logical expressions in function definitions is how we enter a piecewise defined function. Multiply the appropriate function rule by the logical expression for the interval on which it is defined. For example, look at the function definition of $Y_1$ in Figure 21.3. It is the squaring function on the interval $0 < x \leq 1$ and zero otherwise. You would add a piece for each interval of definition of your piecewise defined function.

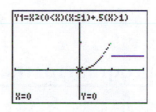

*Figure 21.3 A piece-wise defined function.*

## The triangle wave function

A complete period of the triangle wave function starts at 0, increases linearly to a value of one at $x = 1/2$, then decreases linearly back to zero at $x = 1$. The two pieces of this function can be added together to make a single function definition as shown in Figure 21.4. We could make a more complicated definition that would actually repeat periodically, but we only need one period when we derive the Fourier approximation function.

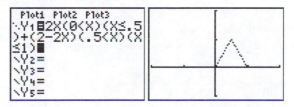

*Figure 21.4 A single period of the triangle wave function.*

# The general formula for the Fourier approximation function

A Fourier approximation function uses the sine and cosine functions to approximate periodic functions. These functions are not polynomials, but we use the polynomial vocabulary to describe them. We use the term *degree* to specify which sine/cosine terms are included and the specific constant multipliers are called the *coefficients*. Here is the definition of the *n*th degree Fourier function for the interval $-\pi \leq x \leq \pi$:

$$F_n(x) = a_0 + a_1 \cos(x) + a_2 \cos(2x) + \ldots + a_n \cos(nx)$$
$$+ b_1 \sin(x) + b_2 \sin(2x) + \ldots + b_n \sin(nx)$$

With this basic structure in mind, we give the generalized definition which includes the general period *b* and the definitions of the coefficients.

$$F_n(x) = a_0 + a_1 \cos((\tfrac{2\pi}{b})x) + a_2 \cos(2(\tfrac{2\pi}{b})x) + \ldots + a_n \cos(n(\tfrac{2\pi}{b})x)$$
$$+ b_1 \sin(x(\tfrac{2\pi}{b})) + b_2 \sin(2(\tfrac{2\pi}{b})x) + \ldots + b_n \sin(n(\tfrac{2\pi}{b})x)$$

$$a_0 = \frac{1}{b} \int_{-b/2}^{b/2} f(x)\,dx$$

$$a_k = \frac{2}{b} \int_{-b/2}^{b/2} f(x) \cos(k(\tfrac{2\pi}{b})x)\,dx \quad \text{for } k > 0$$

$$b_k = \frac{2}{b} \int_{-b/2}^{b/2} f(x) \sin(k(\tfrac{2\pi}{b})x)\,dx \quad \text{for } k > 0$$

## A program for the Fourier approximation function

If there was ever a task calling for programming, this is it. We use a program called FOURIER to do the messy work for us. If possible, find a copy that you can load through the LINK.

### FOURIER program listing

```
ClrHome
Disp "SET:"
Disp "Y₁, WINDOW"
Disp "PUTS:"
Disp "COEFFS IN L₅,L₆"
Disp "FPOLYS IN Y'S."
Disp ""
Disp "Ø=QUIT, 1=SETUP"
Input "2=GRAPH ONLY ",A
ClrHome
If A=Ø
Stop
If A=1:Then
Func
7→dim(L₅)
7→dim(L₆)
Input "PERIOD? ",B
2π/B→C
Disp "WAIT..."
For(K,1,7)
(2/B)fnInt(Y₁*cos(K*C*X),X,Ø,B)→L₅(K)
(2/B)fnInt(Y₁*sin(K*C*X),X,Ø,B)→L₆(K)
End
"(1/B)fnInt(Y₁,X,Ø,B)+L₅(1)cos(C*X)+L₆(1)sin(C*X)"→Y₂
"Y₂+L₅(2)cos(2C*X)+L₆(2)sin(2C*X)"→Y₃
"Y₃+L₅(3)cos(3C*X)+L₆(3)sin(3C*X)"→Y₄
"Y₄+L₅(4)cos(4C*X)+L₆(4)sin(4C*X)"→Y₅
"Y₅+L₅(5)cos(5C*X)+L₆(5)sin(5C*X)"→Y₆
"Y₆+L₅(6)cos(6C*X)+L₆(6)sin(6C*X)"→Y₇
"Y₇+L₅(7)cos(7C*X)+L₆(7)sin(7C*X)"→Y₈
End
ClrHome
Input "GRAPH POLY (1-7)? ",A
If A<1 or A>7
Stop
FnOff :FnOn 1,A+1
DispGraph
Text(1,3,A)
```

# Setting up and using FOURIER

Enter the function in Y₁ with a definition that spans one complete period starting at zero. (This is why we only needed one period of the square wave and triangle wave functions.) Find a nice window for Y₁ that shows several periods on each side of the origin. Check the lists L₅ and L₆ to be sure they don't contain data you want — they will be erased. All function definitions are erased as well and used to store the 'polynomials' of the various orders.

We use the FOURIER program to create and graph the Fourier approximation function of the square wave function. This sequence is shown in Figure 21.5. After you set up Y₁ and a window, start the program from a fresh line on the home screen by selecting FOURIER from the PRGM menu. The opening screen is just a reminder that Y₁ should be the original function. Choose 1=SETUP to calculate the coefficients and store the functions. Next you are prompted for the period and then it takes about one minute to calculate and store the coefficients in L₅ and L₆. The final prompt is a choice of the order of polynomial to graph. When the graph is drawn (this also takes a minute), and you wish to see another graph, press CLEAR, ENTER and it shows the opening screen where you choose 2=GRAPH ONLY for graphs of other degrees.

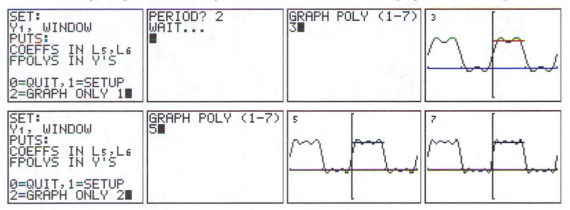

*Figure 21.5 Setting up and using the FOURIER program on the square wave function.*

### The graphs produced by FOURIER

In Figure 21.5 you see three different graphs approximating the square wave function. They are of orders 3, 5, 7, and they get better as the order increases. Remember that it is only required to define and graph one complete period of the original function, yet the polynomials are defined on all real numbers.

### Fitting the triangle wave function

We close with an approximation of the triangle wave function as defined in Figure 21.4. It has a period of 1 and the polynomials are a close fit, even with degree 4. The triangle wave function has a basic shape closer to the sine curve than the square function and thus it makes sense that the fit is so good by degree 4.

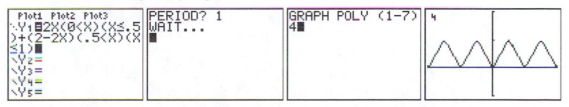

*Figure 21.6 A Fourier function of degree 4 fits the triangle wave function very closely.*

**Notes:**

# CHAPTER TWENTY TWO

# DIFFERENTIAL EQUATIONS AND SLOPE FIELDS

Often a situation occurs where we know about a rate, but the original function is not explicitly known. For example, we saw how to antidifferentiate the equation $\frac{dQ}{dt} = t$ and solve for $Q$ to find $Q = t^2 + C$. This is a simple case, but the rate might depend on $Q$; then the antidifferentiation is not so straightforward. Equations, such as these that involve derivatives, are called differential equations.

Most rates are expressed in terms of time, so $t$ is most often the independent variable. The dependent variable is often a quantity, so $t$ and $Q$ are commonly the variables of differential equations. It is also quite common to use the standard $x$ and $y$ variables, with a prime denoting a derivative. For example, the above differential equation is also written more succinctly as $y' = x$.

## A word about solving differential equations

This calculator solves equations numerically; they cannot tell you the solution as a formula, but they give you a table or a graph of the solution. Thus the calculator is most useful in checking if your analytic solution agrees with the numerically derived solution graph or table. The numerically derived solution is also helpful in situations where you guess-and-check to find an analytic solution. Looking at a graph helps you guess the type of solution function.

### General vs. particular solutions

Analytic solutions are written with the constant $C$. This means that there is a whole set of solutions which differ by a constant. To graph this whole set would fill the screen. So a slope field graph is used to selectively show local slope behavior. The main purpose of this chapter is to introduce this tool.

However, we often know specific conditions that determine a single solution, called a particular solution. Particular solutions are graphed as curves. Graphing these solution curves is the topic of the next chapter. It is common to draw a slope field graph to show the general solution and then superimpose a particular solution curve starting at some initial value.

### Discrete vs. continuous representation

Some calculators, such as the TI-89, are blessed with a differential equation mode, where the differential equations are entered directly and graphed. This model does not have such a mode, but their sequence mode (MODE, SEQ) allows entering and graphing difference equations — a kind of discrete differential equation. These two approaches, the discrete and continuous, both offer insight into the topic. However, it is noted that using the TI-84/83 for differential equations is like using a car as a moving van.

# The SEQ mode

To use the TI sequence mode, press MODE and set the fourth line to SEQ. (Don't forget to press ENTER to change the setting.) Pressing the Y= key shows the three sequence functions u, v and w. The independent variable now is *n* and the X,T,θ,*n* key pastes *n* instead of an X.

> *Tip:* The SEQ mode should be distinguished from the seq command, which was used in earlier chapters.

### A simple sequence for modeling folding

Some situations are modeled clearly and efficiently by defining a sequence recursively, that is, relating a term to its predecessor. If a paper is folded in half over and over; we use the function $y = 2^x$ to determine the number thicknesses after *x* folds. This is a continuous function that does the job. But in fact, folding is discrete. If $u(n)$ is the thickness (in sheets) after *n* folds, then

$$u(n) = 2*u(n-1) \text{ where } u(0)=1$$

This relationship is very clear: double the previous value (after starting at 1). In the sequence mode, this is exactly how we enter the function and its initial value. To enter u, use the 2nd_u key above the 7 key. Like Y functions, the sequence values are displayed using a table. See Figure 22.1. When you enter a number for u(*n*Min) the value is automatically enclosed in set brackets. This quirk is explained in the next chapter.

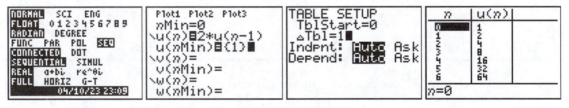

Figure 22.1 Entering the doubling function in SEQ mode and using TABLE to view values.

> *Tip:* You can easily distinguish the sequence functions u, v, w because they are defined using lower case letters. (They are 2nd mode characters above 7, 8, 9 on the keypad.) They are not the same as the upper case U, V, W which are variables.

# The discrete learning curve using sequence mode

One theory of learning is that the more you know, the slower you learn. Let *y* be the percentage of a task we know. The learning rate is then $y'$ and we assume $y' = 100 - y$. The time unit for this rate is the standard five day work week. The slowing of learning takes place immediately and continuously, but we start by looking at a discrete model where we assume the same rate all day.

Consider a new employee who knows 0% of a task at time 0, Monday morning. She learns at the rate $y' = (100-0)\%$ during the first day. A manager, not versed in learning theory, divides the training for the task into five equal parts. On Monday, the part of the task the employee learns is

$$y' * (1/5) = 100\%(0.2) = 20\%$$

At this rate, she would have the task entirely mastered in a week. But Tuesday, because she already knows 20% of the task, her learning slows to

$$y' = 100\% - 20\% = 80\%$$

Therefore, on Tuesday the part of the task she learns is an additional

$$y' * (1/5) = 80\%(0.2) = 16\%$$

In total, she has learned 20% + 16% = 36% by the end of Tuesday.

### Making a table from a sequence

Figure 22.2 shows how to set up this learning function as a sequence and then display it as a table of daily progress.

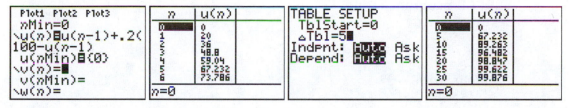

*Figure 22.2 Making a table to show the learning function, then resetting the table increment to see larger values.*

Despite the manager's intention, the employee learns only 67% of the task by the end of the week. By the end of her second week, the new employee knows 89% of the task. According to the model, it is impossible to completely learn the task, but the mastery reaches 99% in the fifth week.

### Graphing a sequence

Graphing a sequence function is similar to graphing a Y function: use the WINDOW key to set up the viewing window. The choices on this screen are now expanded at the top to include four new entries shown in Figure 22.3. These set *n*Min and *n*Max at the start and stop value. The PlotStart value is an integer that starts the graph at some *n* value equal to or beyond *n*Min. The PlotStep value is similar to Xres in the Y function graphing mode; it speeds up graphing if needed.

*Figure 22.3 Graphing the learning function in SEQ mode and using TRACE.*

---

**Tip:**   Setting a window and pressing TRACE graphs and exhibits the function definition, then allows you to use the arrow keys to explore sequence function values.

---

**Tip:**   Because each entry depends on the previous one, it is faster to trace forward. Tracing backwards causes more calculation delay.

# A program for slope fields

The SLPFLD program draws the slope field for a differential equation of the form $y' = F(x, y)$. The indentation in the listing is to help make the program more readable; each line must actually be entered at the left margin, without leading spaces. The program divides the screen into $8*12 = 96$ little boxes and finds a slope line through the middle of each box. The number of slope lines is adjusted by resetting the variables A and B. The 0.3 value that occurs several times is used to shrink the line length within each box. This prevents lines from reaching into neighboring boxes. If you prefer longer lines, change the 0.3 multipliers to 0.5 and use If abs(Z-S)>V in the program line that checks for vertical slope lines.

## SLPFLD program listing

```
ClrDraw
FnOff
8→A
12→B
(Ymax-Ymin)/A→V
(Xmax-Xmin)/B→U
For(I,1,A)
 For(J,1,B)
  Ymin+V*(I-1)+V/2→Y
  Xmin+U*(J-1)+U/2→X
  Y₁→M
  Y-M*.3U→S
  Y+M*.3U→Z
  X-.3U→P
  X+.3U→Q
  If abs(Z-S)>.6V:Then
   Y-.3V→S
   Y+.3V→Z
   .3(S-Y)/M+X→P
   .3(Z-Y)/M+X→Q
  End
  Line(P,S,Q,Z)
 End
End
```

## Setting up and using SLPFLD

Before using the SLPFLD program showing slope lines of a differential equation of the form $y' = F(x, y)$, place the $F(x, y)$ side of the equation in Y₁. This means that Y₁ can be an expression using the variables X and Y (use ALPHA_Y for the latter). It is still the case that Y₁ functions require X as the variable for normal graphing; but the SLPFLD program gets away with its graphing by not using the GRAPH mode. Set what you hope is an appropriate window, but you must use the SLPFLD program to see if it is a nice window; pressing the GRAPH key usually results in an error. A typical setup and graph are shown in Figure 22.4.

The graph the slope lines of learning function, $dy/dx = 100-y$, in Figure 22.4 shows that the values converge to 100 as X gets larger — this is independent of where X starts. (X is in weeks.) A curve starting at the origin would be the one shown in Figure 22.3. We now show several examples graphing slope lines for various differential equations.

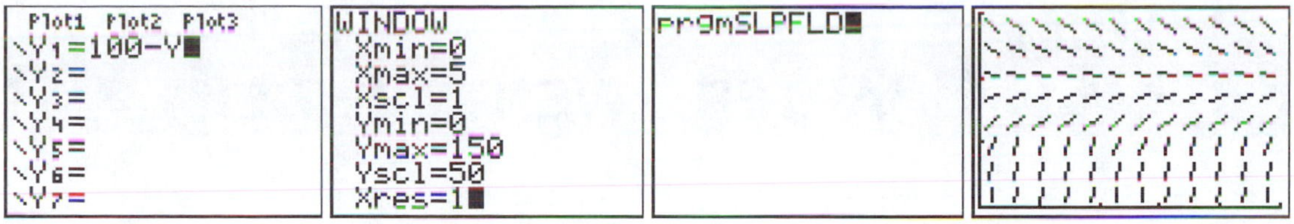

```
Plot1 Plot2 Plot3
\Y1=100-Y■
\Y2=
\Y3=
\Y4=
\Y5=
\Y6=
\Y7=
```

```
WINDOW
Xmin=0
Xmax=5
Xscl=1
Ymin=0
Ymax=150
Yscl=50
Xres=1■
```

```
prgmSLPFLD■
```

*Figure 22.4 Slope field for the learning function.*

## Slope fields examples for differential equations

### Slope field for $y' = 2x$

Use a ZDecimal window for the next four examples. The solution graphs are parabolas of the form $y = x^2 + C$.

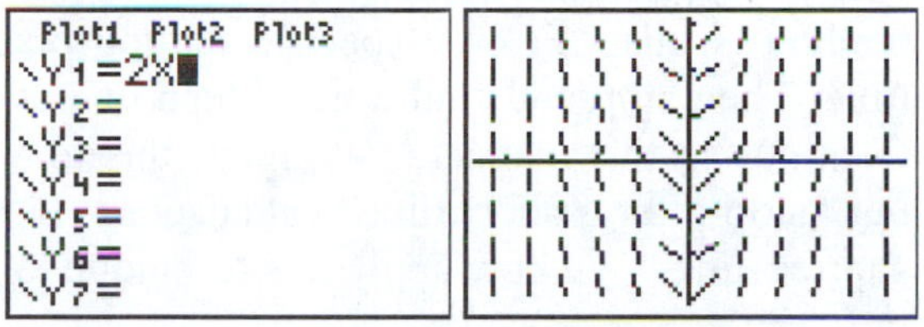

```
Plot1 Plot2 Plot3
\Y1=2X■
\Y2=
\Y3=
\Y4=
\Y5=
\Y6=
\Y7=
```

### Slope field for $y' = y$

The solutions are of the form $y = e^x + C$.

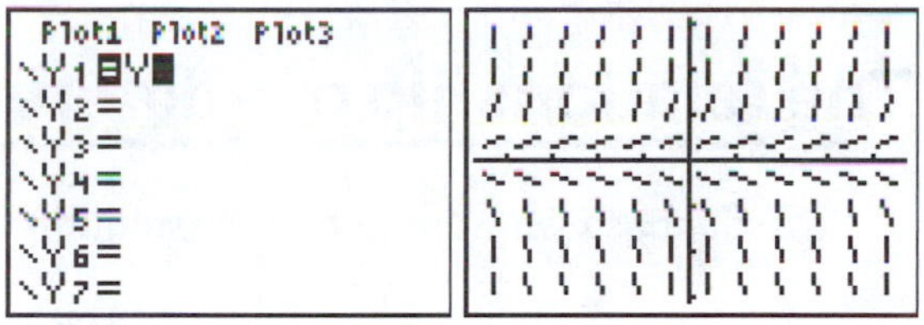

```
Plot1 Plot2 Plot3
\Y1=Y■
\Y2=
\Y3=
\Y4=
\Y5=
\Y6=
\Y7=
```

### Slope field for $y' = x / y$

The solutions have hyperbolas as graphs.

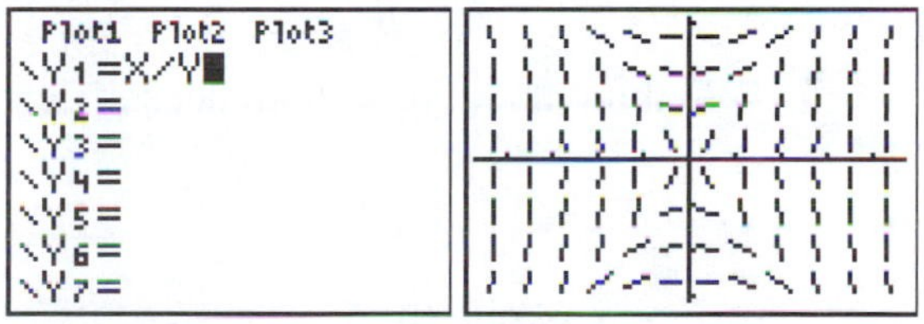

```
Plot1 Plot2 Plot3
\Y1=X/Y■
\Y2=
\Y3=
\Y4=
\Y5=
\Y6=
\Y7=
```

### Slope field for $y' = -x / y$

The solutions are circles centered at the origin. By implicit differentiation, we know that this differential equation is obtained from $x^2 + y^2 = a$ where $a > 0$.

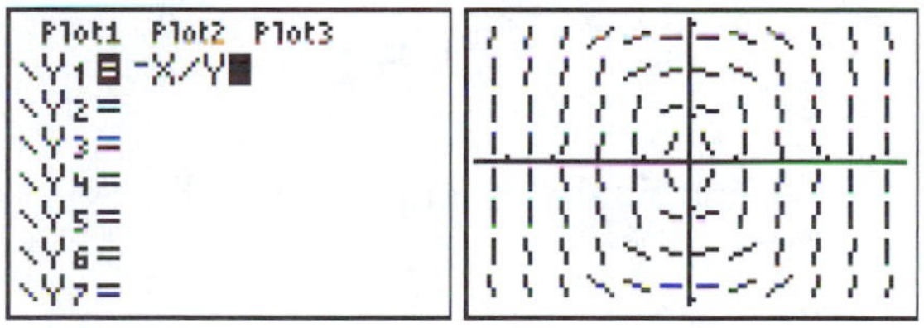

```
Plot1 Plot2 Plot3
\Y1=-X/Y■
\Y2=
\Y3=
\Y4=
\Y5=
\Y6=
\Y7=
```

### Slope field $y' = (-y + xy) / (x - xy)$

The predator-prey model shows a slope field that relates two interacting populations A particular solution is an oval curve, as the populations are cyclical, unless we are at (1,1), a fixed point. Window $0 \leq x \leq 3$ and $0 \leq y \leq 3$. This equation is discussed in Chapter 25.

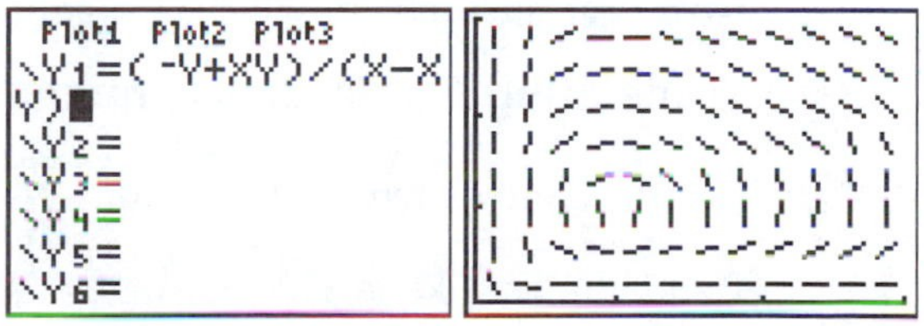

```
Plot1 Plot2 Plot3
\Y1=(-Y+XY)/(X-X
Y)■
\Y2=
\Y3=
\Y4=
\Y5=
\Y6=
```

# CHAPTER TWENTY THREE

# EULER'S METHOD

In the last chapter we discussed how to graph particular solutions to differential equations when we explicitly knew the function of the solution. When we don't know the explicit formula, we can still graph an approximation curve using a technique called Euler's method. The general idea of Euler's method is this: "From a specific starting place, follow the slope line through this point for a short time. Then stop and find a new slope line at my current point and follow that line for a short time. Continue in this manner." We relate this to a differential equation and use the sequence mode to do our "stop after a short time" calculations. If better accuracy is needed, then we stop for shorter and shorter times. We now apply this technique to particular solutions of several differential equations.

## The relationship of a differential equation to a sequence

Suppose we want to find the solution of the differential equation

$$\frac{dy}{dx} = F(x, y) \text{, starting at the point } (x_0, y_0)$$

We think about this continuous differential equation in a discrete way and write

$$\frac{\Delta y}{\Delta x} = F(x, y)$$

or, more explicitly,

$$\frac{y_n - y_{n-1}}{\Delta x} = F(x_{n-1}, y_{n-1}) \text{ with } \Delta x = x_n - x_{n-1}$$

and finally, multiply both sides by $\Delta x$ and rearrange in a form we can use,

$$y_n = y_{n-1} + F(x_{n-1}, y_{n-1})\Delta x \qquad x_n = x_{n-1} + \Delta x \qquad (*)$$

Now you see how knowing $(x_0, y_0)$ allows us to find $y_1$.

### Translating a differential equation to a difference equation

Consider the differential equation $\frac{dy}{dx} = y$ starting at (0,1). With $\Delta x = 0.1$, we set u to act like x and v to act like y. This translates the (*) equations into

$$\begin{aligned} u_n &= u_{n-1} + 0.1 \\ v_n &= v_{n-1} + v_{n-1}(0.1) \end{aligned} \qquad \text{starting at } u_0 = 0 \text{ and } v_0 = 1$$

---

***Tip:*** When entering definitions for u and v, it is safer to enter equation exactly as given and not simplify with algebra.

# Using sequence functions

### Table and graph setup for $y' = y$, starting at (0,1)

First set MODE, SEQ and then press Y= to define any of the three sequence functions u, v, w. Enter the equations as shown in Figure 23.1. Use TBLSET to prepare a simple table starting at $n = 0$. In the last frame of Figure 23.1 you see that $n$ is displayed in the far left column and that the u($n$) and v($n$) columns are the $x$ and $y$ approximations. Since we know $y = e^x$ is the particular solution to this differential equation, we can check and see that the approximation v(6) = 1.7716 is a little below the true value of $y = e^{u(6)} = e^{0.6} \approx 1.8221$.

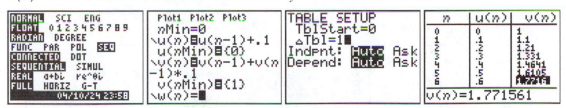

Figure 23.1 *A table of approximations to a particular solution of the differential equation $y' = y$.*

> *Tip:* Remember that in SEQ mode, the $n$ is typed by using the X,T,θ,$n$ key.

### The 2nd_FORMAT UV axes setup

When dealing with functions, there are two variables and the assignment of axes is straightforward. But in SEQ mode, there are four potential variables and therefore several choices for what the axes represent. The default axes assignment (TIME) graphs u or v or w (whichever is selected) on the vertical axis and n on the horizontal axis. Since u corresponds to $x$ we want these values to be on the horizontal axis. The 2nd_FORMAT menu to change the axes assignment from Time to uv which graphs u on the $x$-axis and v on the $y$-axis. We also make adjustments to the window settings, as shown in Figure 23.2. Notice that the additional settings mean that we must arrow down to see and change the Y parameters.

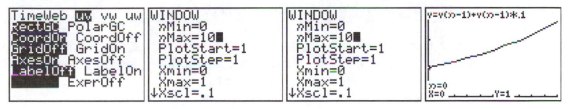

Figure 23.2 *Using 2nd_FORMAT uv to graph the solution of the differential equation $y' = y$.*

> *Tip:* In SEQ mode, TRACE shows $n$, u and v values. The up/down arrows allows you to exhibit either the u or v definition.

# Euler's method for *y′* = -*x* / *y* starting at (0,1)

Consider the differential equation $\dfrac{dy}{dx} = \dfrac{-x}{y}$ starting at (0,1). We investigate the effect of using a smaller value of $\Delta x$ by graphing the solution on the interval $0 \le x \le 1$ with $\Delta x = 0.01$. We set u to act like *x* and v to act like *y*. This translates to the difference equations:

$$u_n = u_{n-1} + 0.01$$
$$v_n = v_{n-1} + (-u_{n-1}/v_{n-1})(0.01)$$

starting at $u_0 = 0$ and $v_0 = 1$.

### Graphing *y′* = -*x* / *y* using Euler's method

```
Plot1 Plot2 Plot3
\u(n)🔲u(n-1)+.01
 u(nMin)🔲{0}
\v(n)🔲v(n-1)+(-u
(n-1)/v(n-1))(.0
1)
 v(nMin)🔲{1}■
```
```
WINDOW
 nMin=0
 nMax=100■
 PlotStart=1
 PlotStep=1
 Xmin=-2.35
 Xmax=2.35
↓Xscl=1
```
```
WINDOW
↑PlotStep=1
 Xmin=-2.35
 Xmax=2.35
 Xscl=1
 Ymin=-1.55
 Ymax=1.55
 Yscl=1■
```

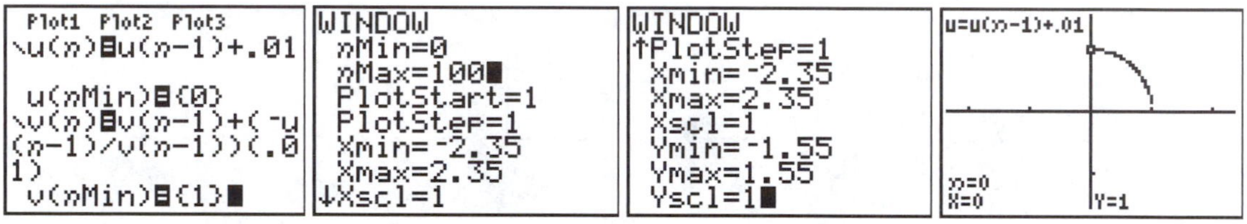

*Figure 23.3 Graphs of y′ = -x/y using Euler's method with $\Delta x = 0.01$.*

Figure 23.3 shows the screens used to set up and graph these difference equations using Euler's method with $\Delta x = 0.01$. Remember that the FORMAT is set to uv as in the previous example. The window settings are half of the standard ZDecimal. This graph looks like part of a circle of radius one, which we know is the actual solution.

> *Tip:*   A quick way to get a nice window is to begin with ZDecimal and improve it using Zoom In or Zoom Out. Under ZOOM MEMORY 4:SetFactors, you can set XFact=2 and YFact=2 for better zoom control.

# Euler gets lost going around a corner

Use Euler's method with caution. As you wander away from your initial point, you may encounter increasing errors. The next example shows that there are some paths where we encounter an infinite slope that will cause problems.

Let's take the previous example and try letting *n* go beyond 100 in hopes that it will sweep through the *x*-axis and continue to follow its circular path. Set *n*Max = 150 and graph again. Figure 23.4 shows the folly as we come to (1,0).

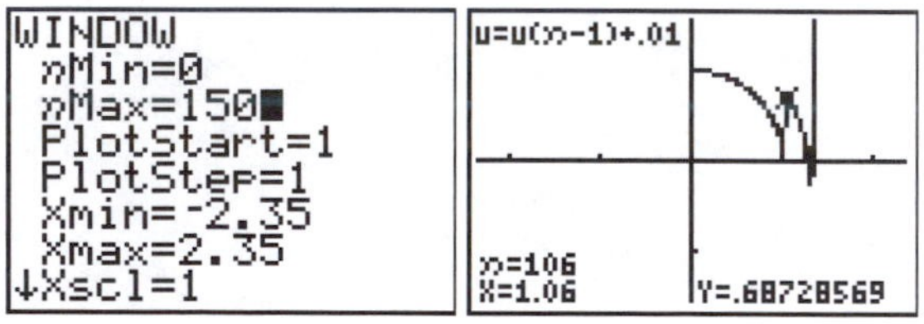

```
WINDOW
 nMin=0
 nMax=150■
 PlotStart=1
 PlotStep=1
 Xmin=-2.35
 Xmax=2.35
↓Xscl=1
```

*Figure 23.4 Euler's method breaks down as the slope is undefined at (1,0).*

To get a circular path in the fourth quadrant, we can start at (0,-1), where the slope is defined. In Figure 23.5 the solution curve moves along the correct path.

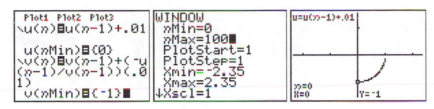

*Figure 23.5 Euler's method finds the way by going from (0,-1).*

## A note about the Fibonacci sequence

You may have wondered why there is the list notation { } placed around initial values of the sequence functions. This is because a list of initial values is used in case the references go back further than the previous term. For example, a term of the Fibonacci sequence is defined as the sum of the previous two terms: $u(n) = u(n-1) + u(n-2)$. Thus we need two initial values to get started: $u(1) = 1$, $u(2) = 1$. We see in Figure 23.6 the definition, initial values, and seven terms of the Fibonacci sequence.

> *Tip:*    If the FORMAT is uv, then the TABLE command calculates and displays the sequence values for both u and v. If v has no definition it creates an error (even if v is not selected.) Change FORMAT to the Time default to correct this problem.

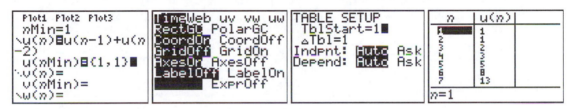

*Figure 23.6 The definition and initial values for the Fibonacci sequence, to get a sequence table giving the first seven terms.*

# CHAPTER TWENTY FOUR

# THE LOGISTIC POPULATION MODEL

In this chapter we derive a logistic equation to model the population growth of the United States The exposition here is a brief introduction of the TI statistical capabilities — a whole new side of the calculator.

## Entering US population data 1790-1940

You may wonder why we don't list the most current census data, the answer is that we want to use these earlier years to make a population prediction equation and then compare its prediction with the actual more current data.

| Year | 1790 | 1800 | 1810 | 1820 | 1830 | 1840 | 1850 |
|---|---|---|---|---|---|---|---|
| Years since 1790 | 0 | 10 | 20 | 30 | 40 | 50 | 60 |
| Population ($\times 10^6$) | 3.9 | 5.3 | 7.2 | 9.6 | 12.9 | 17.1 | 23.1 |

| 1860 | 1870 | 1880 | 1890 | 1900 | 1910 | 1920 | 1930 | 1940 |
|---|---|---|---|---|---|---|---|---|
| 70 | 80 | 90 | 100 | 110 | 120 | 130 | 140 | 150 |
| 31.4 | 38.6 | 50.2 | 62.9 | 76.0 | 92.0 | 105.7 | 122.8 | 131.7 |

*Table 24.1 US Census Data 1790 – 1940.*

The first step is to enter data given in Table 24.1 from the US Census Bureau into two lists. Typically, annual data is not indexed by the year itself, but by years since a base year. Our base year is 1790 and the populations is rounded to the nearest tenth of a million. Use STAT, EDIT to enter the data into $L_1$ and $L_2$. You may prefer to start by enter $L_1$ using the seq() command from the home screen as follows:

$$\text{seq}(1\emptyset * I, I, \emptyset, 15) \to L_1$$

This quickly enters the whole list in $L_1$. Next use STAT EDIT, ENTER and fill in $L_2$. The data screens are shown in Figure 24.1.

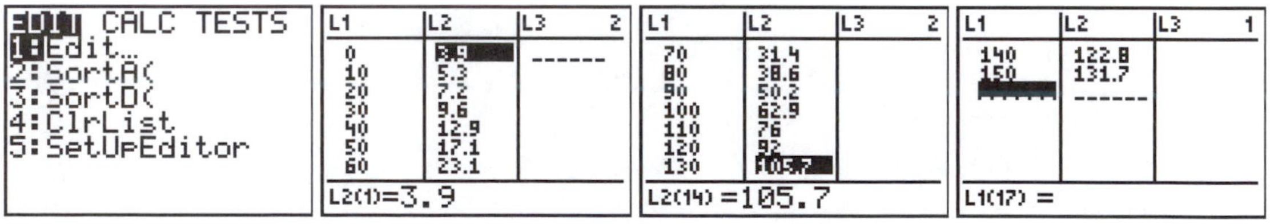

*Figure 24.1 Use STAT EDIT to enter US population data 1790 – 1940.*

# Estimating the relative growth rates: $P' / P$

We want to find the relative annual growth rate, but we settle for an approximation that uses data from the present and next decade:

$$\frac{1}{P}\frac{dP}{dt} \approx \frac{1}{P_i}\frac{P_i - P_{i-1}}{10}$$

Figure 24.2 shows using the `seq()` command to do these calculations and store them in $L_3$.. Looking at $L_3$ values in frame two, we see that the rate shown is more or less constant with a relative growth rate of 3.5% over each 10 years. We find the exact average using `mean` from the 2nd_LIST, MATH menu and then use `Solver` to find the continuous growth rate needed to write the exponential model equation in base e.

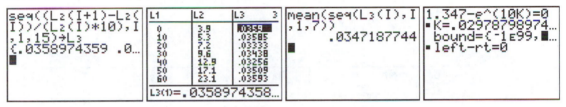

Figure 24.2 Using the `seq()` command to estimate the relative growth rate and store it in list $L_3$. Last frame: the annual continuous growth rate is found using `Solver`.

## Modeling population growth with a simple exponential function

Using the above results, a naïve model builder might say the population starts at 3.9 (million) and grows at a 2.98% continuous rate so that the following simple exponential model is used:

$$y = 3.9e^{0.0298x}$$

In Figure 24.3 we see the folly of this approach as the population for later years did not rise as fast as predicted. The graph of the actual data is shown as a decade step function given by $Y_2$. The mechanics of this function are explained in the following tip.

> **Tip:** A little programming trick was used to graph the list $L_2$. The first value, $L_2(1)$ is graphed for $0 \leq X < 10$ since $0=\text{int}(X/10)$. Likewise, each decade is a constant from $L_2$.

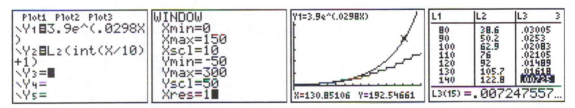

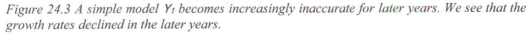

Figure 24.3 A simple model $Y_1$ becomes increasingly inaccurate for later years. We see that the growth rates declined in the later years.

By returning to our table of estimated relative growth rates we see that while they were relatively constant from 1790 to 1860, the growth rate slows from1860 to 1930. This means that our simple exponential model needs to be replaced.

# A scatter plot

One of the remarkable features of a TI calculator is its statistical capability. We make only a brief use of it here (the fuller features can be explored using the *TI Guidebook*.) We create a scatter plot to look for a relationship between years, in $L_1$, and the corresponding population data, in $L_2$.

The 2nd_STAT PLOT key displays a STAT PLOTS screen with the setup status of the three stat plots. After selecting the plot, the setup window allows you to set the following options:

> On Off same effect as the = toggle On Off for Y functions
>
> Type: the first icon indicates scatter plot
>
> XList:, YList: list assigned to each axis
>
> Mark: graphic symbol for the plot points

*In Figure 24.4  Plot1 selected and graphed as a scatter plot for $L_1$ against $L_2$ with a box marker.*

In Figure 24.4, Plot1 has been selected, turned on, and selected as a scatter plot for $L_1$ and $L_2$, with a box marker. Then we set an appropriate window with the handy ZOOM ZoomStat. The scatter plot shows the slowing of the growth rate and for suggests a model called the logistic model.

> *Tip:*    The status of the three plots as On or Off is also shown at the top of the Y= edit screen: the plots that are on are highlighted. The status can be toggled directly from the Y= screen by highlighting the plot name and pressing ENTER.

> *Tip:*    A frequent annoyance is that a Y function is left selected from a previous use and it shows up on the graph screen. (Even if it is outside the window, it still slows the graphing.) Check the Y= screen for proper selections before graphing.

# Finding a regression line to fit the data

Another statistical feature of the TI calculator is its ability to find regression equations (models) of various kinds. The word regression is a statistics term, but for our purposes we can think of it as meaning a *best fit* type of equation. The most common is the regression line equation, giving a line that comes closest to graphically fitting our data. We also have exponential regression, giving an exponential equation that best fits the data. Our scatter plot in Figure 24.4 has a logistic curve look, so we fit it using a logistic regression equation.

Figure 24.5 shows the use of STAT, CALC, D:Logistic on $L_1$ and $L_2$ to find the equation. Some explanation is needed for this huge number of screens for a simple task. The first three frames show the calculation of the logistic regression equation coefficients. At that point the coefficients could be rounded and entered to $Y_1$ by hand for graphing. However there is a sequence of steps that paste in the equation at full accuracy. Use VARS, Statistics, EQ, RegEQ to make the transfer. If you often use regression equations this sequence becomes more natural.

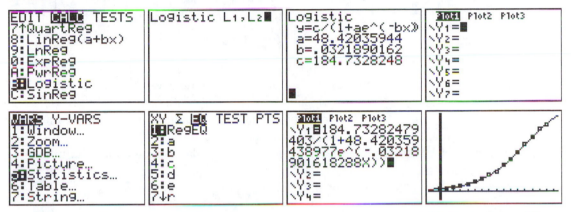

Figure 24.5 *Finding and graphing the logistic regression equation. Using* VARS Statistics,EQ *to paste in full accuracy of coefficients.*

Our logistic model equation is

$$P = \frac{185}{1 + 48e^{-0.0322t}}.$$

This equation is not a very good model for the current population because of the unexpected baby boom as shown in Figure 24.6.

Figure 24.6 *Adding new data and graphing the logistic regression equation.*

It is left to the reader to recalculate a regression equation using the data through the year 2000 and to graph a current projection for US population growth.

Tip:    Using the proper regression equation to model a situation is important. The various STAT CALC regression options allow you to try out different models on the same data to look for a good equation.

# CHAPTER TWENTY FIVE

# SYSTEMS OF EQUATIONS AND THE PHASE PLANE

In this chapter we look at examples of systems of differential equations where the independent variable is time. These equations are graphed to show their relationship to each other. Points on the graph are called the phase trajectory or orbit in the phase plane. Two popular examples of using systems of differential equations are the S-I-R model the predator-prey model.

For example, in a predator-prey model we graph the predator population with time on the $x$-axis and then also graph the prey population with time on the $x$-axis. We can also create a phase plane graph with predator population on the $y$-axis and prey population on the $x$-axis.

## The S-I-R model

The S-I-R stands for Susceptible - Infected - Recovered, so you can guess that this is used to model epidemics. The population is divided into the three groups and people move from S to I to R, or they just stay in S. The three rates in terms of time are

$$\frac{dR}{dt} = bI$$          the recovery rate depends upon the number infected

$$\frac{dS}{dt} = -aSI$$          the susceptible rate is negative and depends on both the number of infected and susceptible

$$\frac{dI}{dt} = aSI - bI$$          the infected rate is the negative of the sum of the other two rates.

We concentrate on the last two rates, because knowing any two of the quantities $S$, $I$, $R$ automatically determines the third.

### The boarding school epidemic

The following simple example is used to illustrate the S-I-R model. There were 762 students in a boarding school and one returned from vacation with the flu. Two more students became sick the second day. From this we know

$$S = 762 \text{ and } I = 1 \text{ and } \frac{dS}{dt} = -2.$$

We use this to find $a = -0.0026$ from

$$\frac{dS}{dt} = -aSI \text{ , where } (-2) = a(762)(1)$$

Since this flu lasts for a day or two, we assume that half of the sick will get well each day, thus, $b = -0.5$.

## Time plots for the model

In Figure 25.1 set SEQ mode, 2nd_FORMAT Time to define the three difference equations using $\Delta x = 0.1$. We select v and w to view $S$ and $I$ (against time). We look at the epidemic for 20 days, which means we set the window with Xmax=200, since $20/\Delta x = 200$. There are just under 800 students, so we set Ymax=800. We see that the susceptible are in a constant decline and that the number of infected peaks in the sixth day.

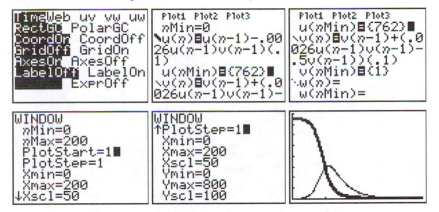

*Figure 25.1 The graphs of S and I over time.*

## Phase plots for the model

In Figure 25.2 we look at the *SI* phase plane. The same definitions of u and v remain from the previous plot, but we now change the FORMAT to uv. We use Xmax=800 and Ymax=400 in the WINDOW. Watching the screen as it graphs, notice that the graph is drawn from right to left, since the first point is $(S, I) = (762, 1)$. Using TRACE on this phase plot feels backward because the right arrow key increases $n$ and causes the trace cursor to move left. If you press the left arrow key at the start, then nothing happens. We see the peak of this curve has a horizontal axis value $v \approx 200$; this is called the threshold value. You can use TRACE to find the peak more accurately at $(191, 313)$, where $n = 55$. We will see an interesting fact about the threshold value when we look at the slope field.

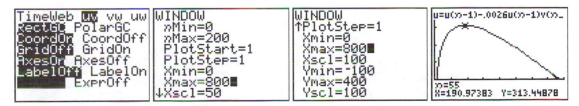

*Figure 25.2 The SI phase plane.*

**Tip:**    The TRACE cursor moves through the points connected to the n values when the right arrow is pressed; the cursor movement is not necessarily from left to right.

### Slope field for the S-I-R model

We use the chain rule in the form of

$$\frac{dI}{dt} = \frac{dI}{dS}\frac{dS}{dt} \quad \text{to get} \quad \frac{dI}{dS} = \frac{dI/dt}{dS/dt}$$

We simplify our original equations in this way to get

$$\frac{dI}{dS} = \frac{aSI - bI}{-aSI} = -1 + \frac{b/a}{S}$$

Thus in this case (with rounding)

$$\frac{dI}{dS} = -1 + 192/S$$

In Figure 25.3 we have used the SLPFLD program to graph the slope lines. Of particular interest is that along any particular solution, the peak (threshold value) is at the same value ($S$=192) on the horizontal S-axis. But if $S$ is greater than 192 then the $I$ values decrease immediately.

*Figure 25.3 The SLPFLD program graphs the slope lines. The threshold value of S is the same for all solutions.*

# Predator-prey model

We start with the well-known *Lotka-Volterra* equations where predators are $r$, the number of robins, and prey are $w$, the number of worms. There are four parameters: $a$, $b$, $c$, and $k$.

$$\frac{dw}{dt} = aw - cwr \quad \text{and} \quad \frac{dr}{dt} = -br + kwr$$

We simplify by setting all four parameters to be 1 and have

$$\frac{dw}{dt} = w - wr \quad \text{and} \quad \frac{dr}{dt} = -r + wr$$

We define the predators as v and the prey as u. For graphing by Euler's method, we use $\Delta t = 0.01$, and our sequence equations are

$$u_n = u_{n-1} + (u_{n-1} - u_{n-1}v_{n-1})\Delta t \quad \text{and} \quad v_n = v_{n-1} + (-v_{n-1} + u_{n-1}v_{n-1})\Delta t.$$

From the two equations in the system, using the chain rule, we obtain the single equation

$$\frac{dr}{dw} = \frac{-r + wr}{w - wr}$$

to be used for the graph the slope fields.

## Slope field for the predator-prey model

The slope field of this differential equation might look familiar; it was drawn as the last example in Chapter 22. In Figure 25.5, we see that the particular solutions are oval curves and that there is an equilibrium point at (1, 1). At these values, the two populations theoretically remain stable and do not have the kind of oval cycle that we see if we start from some other point.

*Figure 25.4 Slope curves of a predator-prey model.*

## Phase plots for the predator-prey model

We set MODE SEQ and FORMAT to uv and enter the sequence equations as shown in 25.5. We guess that a complete cycle has taken place after about *n*Max=630. (The slope field graph looked circular and 630 is approximately $2\pi/\Delta x$.) This comes close to giving us one cycle. Remember when tracing values to press the right arrow key even though the trace cursor at first moves left; you are moving up in *n* which may correspond to moving right or left in *x*.

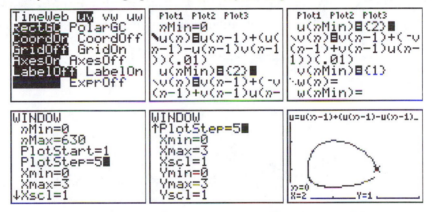

*Figure 25.5 Almost a complete cycle of the predator-prey model.*

## Time plots for the predator-prey model

In Figure 25.6 the individual predator and prey differential equations with respect to time are entered and 2nd_FORMAT is set to Time mode. Although initially you may have no idea how to set the window, a little trial-and-error can lead to the window shown in Figure 25.6. We made it wide enough to show the periodic nature of the two graphs.

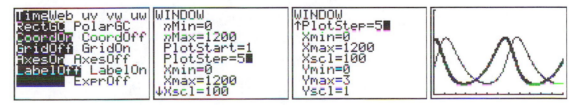

*Figure 25.6 The periodic nature of the predator and prey populations.*

# CHAPTER TWENTY SIX

# SECOND-ORDER DIFFERENTIAL EQUATIONS

A second-order differential equation has a second derivative involved in the equation expression. Before making the transition from a symbolic equation to necessary calculator definitions, we rewrite these equations by solving for the second derivative in the form $y'' = F(x, y, y')$.

## Euler's method for second-order differential equations

As mentioned before, this calculator does not do symbolic solutions of differential equations, but it is useful for checking a solution by graphic means. In our examples we use difference equations to approximate the graph of the solution to a second-order differential equation. The difference equations use Euler's method to graph a solution, without knowing an analytic function for the solution. We compare this graph to a graph of our analytic solution, which we solved by hand. This gives confidence in our analytic results or warns us that we might have made an error.

Set the function mode to SEQ and recall how the u and v were defined in Chapter 23. We chose $\Delta x$ and set u to act like $x$ and v to act like $y$. This translated to equations of the form

$$x_n = x_{n-1} + \Delta x$$

$$y_n = y_{n-1} + F(x_{n-1}, y_{n-1})\Delta x$$

Now in order to get a second derivative equation, we use the next sequence variable, w, to be the derivative $y'$. Let's see how the definitions come out using this substitution.

$$u_n = u_{n-1} + \Delta x$$

$$v_n = v_{n-1} + w_{n-1}\Delta x$$

and with a little work we find w

$$\frac{dw}{dx} = \frac{d(y')}{dx} = y'' = F(x, y, y') \text{ so}$$

$$\frac{w_n - w_{n-1}}{\Delta x} = F(u_{n-1}, v_{n-1}, w_{n-1}) \text{ or },$$

$$w_n = w_{n-1} + F(u_{n-1}, v_{n-1}, w_{n-1})\Delta x$$

We write the three sequence functions u, v, w. in the above form given $y'' = F(x, y, y')$. We start with a simple example where $F(x, y, y')$ is a constant.

# The second-order equation s″ = -g

The equation, $\dfrac{d^2 s}{dt^2} = -g$ , where $g$ is the constant force of gravity on a falling object, $s$ is the displacement in feet, and $t$ is the time in seconds, is a classic of calculus. We assume $g$ is 32 ft/sec$^2$, the initial velocity is zero ($v_0 = 0$), and the initial distance is zero ($s_0 = 0$). By antidifferentiation we get $s' = -32t$ and antidifferentiating again gives the solution $s = -16t^2$.

> *Tip:*  If we did not know the two simple antiderivatives of this example, we might try to use the calculator to apply numerical integration twice:
>
> Y₁=fnInt(-32,X,0,X) and Y₂=fnInt(Y1,X,0,X)
>
> and graph Y₂ as our solution. This nesting cannot be done and results in an error.

Translating the equation to the TI sequence form, we let u = $t$, v = $s$, and w = $s'$. Figure 26.1 shows the definitions for the u, v, w sequences. A good WINDOW is set with *n*Max=10 so that it graphs $x$ from 0 to 1. We overlay the known solution curve, $s$=-16$t^2$, to compare the two graphs. This requires the 2nd_DRAW DrawF command. We see that there is a graphical difference between the two graphs. However, the close similarity gives us faith that one graph is an approximation of the other. Using $\Delta x = 0.01$ makes the two graphs much closer.

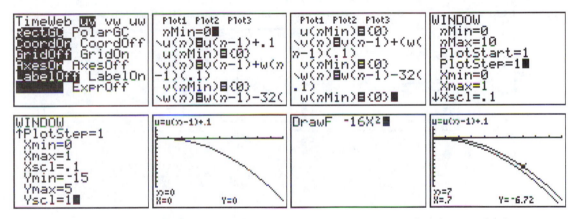

*Figure 26.1 An Euler's method graph of the solution to y″ = -32 compared to the actual solution graph.*

> *Tip:*  Both the Y= and WINDOW screens must be scrolled to see all the entries; they are shown in two screens with some values shown twice.

> *Tip:*  When entering a DrawF function, you need to use ALPHA_X to enter X because the X,T,θ,*n* key in the SEQ mode gives you *n*.

# The second-order equation $s'' + \omega^2 s = 0$

Next we look at a second-order differential equation that depends on only one other variable. A typical example is the equation that describes simple harmonic motion,

$$\frac{d^2 s}{dt^2} = F(s) = -\omega^2 s$$

It is known that the solution to this equation is

$$s(t) = C_1 \cos \omega t - C_2 \sin \omega t$$

### The second-order equation example $y'' = -4y$

For this example let $\omega = 4$ and then set the initial conditions to be $s(0) = 1$ and $s'(0) = -6$. Substituting these initial conditions into the general solution, we find the particular solution has constants $C_1 = 1$ and $C_2 = -3$.

Let's graph an approximation and compare it to this known solution. As in the previous example, we translate the equation to TI sequence form by letting u be $t$, v be $s$, and w be $s'$, and $\Delta x = 0.01$. In Figure 26.2 we enter the definition for w, and reset initial conditions. Since this is a trigonometric solution, we graph $0 \le x \le 2\pi$. (Use ZTrig and alter Xmin) Notice how we set $n$Max, using an integer close to $x/\Delta x$. We that the two graphs are extremely close, but there is more deviation in the second period of the Euler's method graph. We have seen this before: the further out you go, the more inaccurate the method becomes

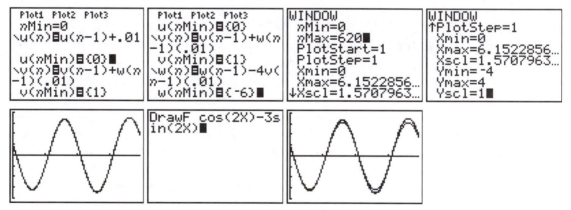

*Figure 26.2 An Euler's method graph of the solution to $y'' = -4y$ with $\Delta x = 0.01$ compared to the actual solution graph.*

***Tip:*** Do not confuse the small omega $\omega$, a traditional constant used in this equation, with w, a TI sequence function.

# The linear second-order equation $y'' + by' + cy = 0$

Equations of this type are called linear second-order equations because, when we isolate the second derivative, it is equal to a linear equation: $y'' = -by' - cy$, where $y$ and $y'$ are the variables and $b$ and $c$ are the coefficients. One application of this general equation is describing the motion of a spring.

### The characteristic equation

The solution of a second-order linear equation hinges on the zeros of the quadratic equation $r^2 + br + c = 0$, namely

$$r = -\tfrac{1}{2}b \pm \tfrac{1}{2}\sqrt{b^2 - 4c}$$

There are three cases, based on the discriminant $b^2 - 4c$, and in each case you solve simultaneous equations for constants $C_1$ and $C_2$ to obtain a general solution equation.

### The overdamped case: $y'' + by' + cy = 0$ with $b^2$ - 4c > 0

The general solution to this case is

$$y(t) = C_1 e^{r_1 t} + C_2 e^{r_2 t}$$

where $r_1$ and $r_2$ are zeros of the characteristic equation and $C_1$ and $C_2$ are found from simultaneous equations created from the initial conditions.

For example, $y'' + 3y' + 2y = 0$ then $b = 3$ and $c = 2$, so the discriminant $3^2 - 4(2) > 0$. Figure 26.3 shows the setup for graphing an Euler's method solution with initial conditions $y = -0.5$ and $y' = 3$ at $t = 0$. Thinking of this as a spring's motion in oil, we see that the spring swings past equilibrium ($y = 0$) and then its motion is damped until it is at rest.

An analytic solution is found by obtaining $r_1 = -1$, $r_2 = -2$, as the roots of the characteristic equation. Solving for $C_1$ and $C_2$ in simultaneous equations created from the initial conditions gives, $C_1 = 2$ and $C_2 = -2.5$. You can check the graphical solution against the symbolic solution by using

```
DrawF 2e^(-X)-2.5e^(-2X)
```

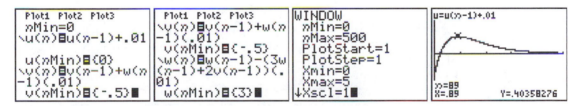

*Figure 26.3 The setup and graph of the overdamped case. (Window $-1 \leq y \leq 1$)*

### The critically damped case: $y'' + by' + cy = 0$ with $b^2$ - 4c = 0

The general solution to this case is

$$y(t) = (C_1 t + C_2)e^{-bt/2}$$

For example, suppose $b = 2$ and $c = 1$, then the discriminant $2^2 - 4(1) = 0$. Figure 26.4 shows the setup for graphing an Euler's method solution with initial conditions $y = -0.5$ and $y' = 3$ at $t = 0$. We see a graph that is similar to the solution of the previous case, but notice that it is not damped as quickly.

An analytic solution is found by solving for $C_1$ and $C_2$ in simultaneous equations created from the initial conditions ($C_1 = 2.5$ and $C_2 = -0.5$). You can check the graphical solution against the symbolic solution by using

```
DrawF (2.5X-0.5)e^(-X)
```

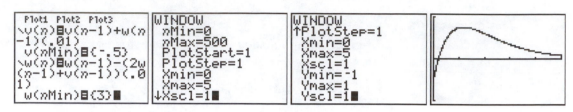

*Figure 26.4 The setup and graph of the critically damped case.*

### The underdamped case: $y'' + by' + cy = 0$ with $b^2 - 4c < 0$

The general solution to this case is

$$y(t) = C_1 e^{\alpha t} \cos \beta t + C_2 e^{\alpha t} \sin \beta t$$

where $r = \alpha \pm i\beta$ are complex zeros of the characteristic equation.

For example, suppose $b = 2$ and $c = 2$, then the discriminant $2^2 - 4(2) < 0$. Figure 26.5 shows the setup for graphing an Euler's method solution when initial conditions $y = 2$ and $y' = 0$ at $t = 0$. We graph the solution from $0 \le x \le 2\pi$ and see that the motion is damped but the graph crosses the equilibrium point at least once in that interval. From the solution equation, we know that this graph is an exponentially damped sine curve. The equation for u is not shown in Figure 26.6, but it is the same as in the previous examples.

An analytic solution is found by obtaining $r_1 = -1 + 2i$, $r_2 = -1 - 2i$, as the roots of the characteristic equation. Solving for $C_1$ and $C_2$ in simultaneous equations created from the initial conditions gives $C_1 = 2$ and $C_2 = 2$. You can check the Euler's method graph solution against the symbolic solution by using

DrawF 2e^(-X)(cos(X)+sin(X))

*Figure 26.5 The setup and graph of the underdamped case.*

# APPENDIX

## Complex numbers

The TI accommodates complex numbers. However, the SOLVER does not accept complex numbers and finds only real solutions. Complex numbers can be entered in either $a + bi$ or $re^{\theta i}$ form. The real and imaginary parts in the form $a + bi$ often extend off the screen after a calculation. You can, of course, scroll over to see the whole number, but it is suggested that you either use MATH 1:▶Frac in hopes of a fraction or reset the Float setting in MODE to show a smaller number of digits (2, 3, or 4 are good settings). See Figures A.1 and A.3 for examples of these two techniques. The MATH CPX menu has a list of functions that can be used with complex numbers.

*Figure A.1 Complex arithmetic examples and the MATH CPX menu.*

### MODE a+bi

You may have seen the $a+bi$ setting in the MODE menu and wondered what it does. Complex number handling is active regardless of whether the $a+bi$ mode has been selected. The difference with this mode on arises when certain expressions evaluate to a complex number. If the complex mode is not set, then the result is an error message. See Figure A.2.

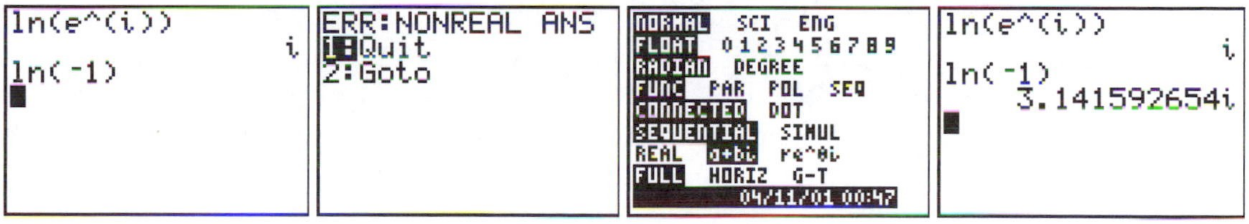

*Figure A.2 In default mode some expressions produce error messages but with MODE a+bi they are evaluated.*

## Polar coordinates in the complex plane

The polar (Pol) graphing mode is based on complex numbers. Each point in the Cartesian rectangular coordinate system has a polar coordinate form $(r,\theta)$ where $r$ is the distance to the origin and $\theta$ is a measure of rotation from the $x$-axis. The related polar form is $re^{\theta i}$. (This notation is also written as $re^{i\theta}$.) An important feature of graphing in the polar form or the parametric form is that the curve need not be a function; the graphs need not pass the vertical line test. Polar graphing is explained fully in its own chapter of the *TI Guidebook*, so we give a very brief presentation here.

## Coordinate conversion

Coordinate conversion tools are found on the MATH CPX menu. You get an error message if you try to convert a real number to a complex one.

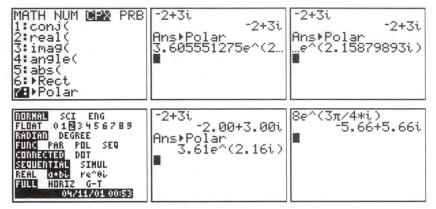

*Figure A.3 Conversion techniques for complex numbers. Changing FLOAT to avoid scrolling and improve readability.*

## Polar graphing

After changing MODE from Func to Pol, the Y= screen shows $r_i$= function definitions. In Figure A.4, a circle of radius 5 is drawn with a ZStandard window. The window ratio is not square, so the graph looks elliptical; this is changed with ZSquare.

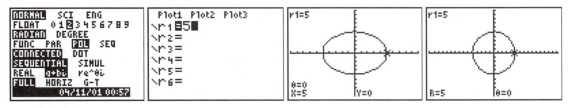

*Figure A.4 Polar graph of a circle in a ZStandard window, use ZSquare to improve the graph.*

In the polar graphing mode, the keypad's variable key is now $\theta$ and equations are defined and graphed with $\theta$ as the variable. In Figure A.5, the equation $r = \theta$ is graphed. It is first traced with 2nd_FORMAT RectGC and then with 2nd_FORMAT PolarGC. Notice in both settings that tracing right increases $\theta$ values which may correspond to moving left or right in $x$.

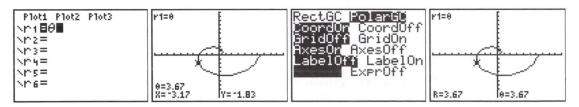

*Figure A.5 The polar equation $r=\theta$ traced with 2nd_FORMAT RectGC and then with 2nd_FORMAT PolarGC.*

# Parametric graphing

The remaining option for graphing is Par, which stands for parametric. In this kind of graph, *x* and *y* are independently defined in terms of a variable *t* (usually thought of as time). The equation for a circle of radius 5 was easy in polar coordinates (Figure A.4). As a parametrically defined curve, a circle is composed of the sine and cosine functions. In Figure A.6 we set MODE to Par, define the equation, and set a standard window. Notice the changes in the Y= screen: equations are now in pairs and the X,T,θ,*n* key gives the variable T. Also, TRACE now includes values of *t*, which satisfy $0 \le t \le 2\pi$ in the standard window.

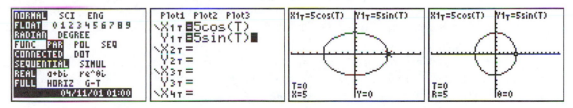

*Figure A.6 Set MODE Par, define parametric equations, and graph in a standard and square window.*

> *Tip:* The only ZOOM feature that changes the t settings is ZStandard. All others ZOOM options reset the window size, but not the t values.

### Writing *y=f(x)* functions in parametric form

Any function $y = f(x)$ can be written in parametric form by setting X₁ᴛ=T and Y₁ᴛ to *f*(T). We see in Figure A.7 that the first graph is restricted to the sine curve with $0 \le t \le 2\pi$. (The standard setting.) In the next graph we have changed Tmin and Tmax so that the graph extends across the window.

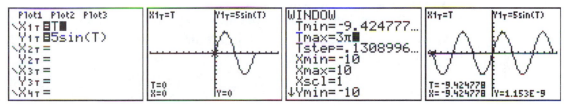

*Figure A.7 Graphing y = f(x) parametrically.*

# Internet address information

The main internet address for Texas Instruments graphing calculator support is

*http://www.education.ti.com/*

At this site you use a search box or dropdown menus to find material of interest. This is the source for the *TI Guidebook*, activities, APPs and product information.

For discussion groups and an extensive downloadable program archive it is far better to use:

*http://www.ticalc.org/*

To go directly to the program archive use:

*http://www.ticalc.org/pub/83plus/basic/math/*

# Linking calculators

The essentials of linking are presented in the *TI Guidebook* and not repeated here. But we include the following tips.

- The end-jack must be pushed firmly into the socket. There is a final click you can feel as it makes the proper connection.
- If you are experiencing difficulty connecting, turn off both calculators, check the connection, and then turn them on and try again. If available, try other cables or calculators.
- When selecting items, the cursor square is hardly visible when the selection arrow is on the same item. As you arrow off of the item, double-check whether or not it has been selected.
- If you are required to drain your calculator memory before entering an exam, use Back Up to keep a copy on some other calculator. Even better is to store it on your computer, which is the next topic.

# Linking to a computer

The TI-84 has a USB connection and comes with a cable that can be connected to a USB port on a computer. The necessary software, TI Connect, is provided on CD and can also be downloaded from the TI website. For the TI-83 and TI-83 Plus a TI-Graph Link™ package must be purchased separately, it is a cable and TI Connect software. There is an older TI-Graph Link™ software program, but the newer TI Connect is more universal.

- This is the best way to back up your work.
- It is the preferred way to write and edit programs.
- You can download and transfer programs from the internet archives.
- It allows you to capture the screen in a form for direct printing or use in a word processor.

# Troubleshooting

### Nothing shows on the screen
- Check the contrast.
- Check ON/OFF button.
- Pull out the batteries and correctly reinsert the batteries.
- As a last resort, remove the backup battery and reinsert all the batteries. Warning: This erases all memory, including programs.

### Nothing shows up on the graph screen except the axes
- Press TRACE to see if a function is defined but outside the window.
- There may be no functions selected.
- The function may be graphed along either axis and need the window reset.
- If there is a busy indicator (a running line) in the upper right corner of the screen, then the TI is still calculating. Press ON if you can't wait.

- If there is a pause indicator (twinkling line) in the upper right corner of the screen, then the TI is paused from a program. Press ENTER to continue.
- If there is a checkerboard cursor, then you have a full memory. You need to delete something; choose some things you no longer need or copy them to your computer.

## Nothing shows up in the table

- You may have the Ask mode set and need to either enter x-values or change to Auto in the 2nd_TBLSET.
- Check to see that a function is selected.

## I get a syntax error screen

The most common errors are

- Parenthesis mismatch. Count and match parentheses carefully.
- Subtraction vs. negative symbol. For example, the subtraction sign cannot be used to enter -10.
- Pasting a command in the wrong place. For example, a program name must be on a fresh line.

## I get an error message

This can cover the widest array of problems. Read the message carefully; it tips you off to the kind of error you are looking for. If you have no idea what caused it, consult the appendix of the *TI Guidebook* for explanations of the error messages.

## I'm getting a result but it is wrong

Check for

- Typing direct letters in place of pasting a command. For example, SIN(X) can be entered using the ALPHA keys, but it is not the same as sin(X) using the SIN key.
- Parenthesis mismatch. Count and match parentheses carefully.
- Subtraction vs. negative symbol. For example, the subtraction sign cannot be used to enter a -10.
- Correct default settings.

## My program won't run

Program errors are difficult to diagnose. Scroll and check your code with PROGRAM EDIT. Add temporary displays and pauses to check the progress and isolate the problem.

**Notes:**

# INDEX